北京市共建项目专项资助

太阳能光伏发电专业英语

古丽米娜 编

中国水利水电出版社
www.waterpub.com.cn
·北京·

内 容 提 要

本书主要介绍太阳能光电转换基本原理、光伏材料与器件基础知识及光伏技术的应用；各种类型太阳电池的光电性能，如晶硅类太阳电池、化合物类太阳电池、有机薄膜类太阳电池、钙钛矿太阳电池以及新型量子点太阳电池等。本书结合太阳能光伏发电的基本转换原理、技术利用方式以及工程应用的进展，引出所需掌握的光伏专业英语核心知识，为进一步提高读者光伏专业英语阅读、写作能力奠定坚实的基础。

本书既可作为光伏专业人员的参考资料，也可作为相关专业的教学用书。

图书在版编目（CIP）数据

太阳能光伏发电专业英语 / 古丽米娜编. -- 北京 : 中国水利水电出版社, 2018.10(2021.3重印)

ISBN 978-7-5170-7008-5

Ⅰ. ①太… Ⅱ. ①古… Ⅲ. ①太阳能发电—英语 Ⅳ. ①TM615

中国版本图书馆CIP数据核字(2018)第231087号

书　　名	**太阳能光伏发电专业英语** TAIYANGNENG GUANGFU FADIAN ZHUANYE YINGYU
作　　者	古丽米娜　编
出版发行	中国水利水电出版社 （北京市海淀区玉渊潭南路1号D座　100038） 网址：www.waterpub.com.cn E-mail：sales@waterpub.com.cn 电话：（010）68367658（营销中心）
经　　售	北京科水图书销售中心（零售） 电话：（010）88383994、63202643、68545874 全国各地新华书店和相关出版物销售网点
排　　版	中国水利水电出版社微机排版中心
印　　刷	清淞永业（天津）印刷有限公司
规　　格	184mm×260mm　16开本　8.5印张　269千字
版　　次	2018年10月第1版　2021年3月第2次印刷
定　　价	**45.00**元

前言
FOREWORD

随着环境状况的日益恶化、化石能源的逐渐枯竭，新能源、新材料的开发及应用已成为全球的首要课题。其中，光伏材料、器件及产业的发展是重点之一。因此，在全球大力发展新能源尤其是光伏产业之际，掌握光伏基础知识及相关信息非常重要，将其作为高等基础教育的一部分也尤为关键。与此同时，英语作为一种重要的全球化交流工具，在国际交流合作中发挥着非常重要的作用。因此，学好光伏发电专业英语，是学生、学者和工程技术人员与国际光伏同行交流，积极获取国际光伏科研信息，掌握国外光伏学科发展动态，参加国际光伏学术交流的前提。为此，本书为高等院校的学生、工程技术人员等提供了所需掌握的一些光伏英语核心知识，希望能对各阶段光伏相关人员专业英语水平的提高有所帮助。

本书共分为8章，第1章主要介绍光伏发电的基础知识；第2～6章介绍各种光伏发电材料以及新型的光伏电池；第7章和第8章介绍光伏发电技术的应用。涵盖了基础知识、应用技术以及新兴的新型光伏发电材料的内容。每章均由专业知识内容、专业词汇、练习题、课后延伸阅读等组成，且每章的课后延伸阅读紧密围绕该章主题，并在此基础上进一步扩展相关学术研究内容，使得阅读者能更广泛地接触学术前沿知识。

本书在帮助学生掌握光伏发电材料、光伏发电技术应用相关知识的基础上，进一步提高学生的专业英语阅读能力，拓展和深化学生对光伏发电技术的认知。本书是为培养学生专业英语应用能力而编写的教材，其特点如下：

（1）针对性强。本书内容贴合光伏发电材料、光伏发电技术应用相关专业知识，内容更全面、更准确。

（2）内容全面。本书各章节内容从光伏发电基本原理、光伏发电材料种类到光伏发电技术应用方式等，从光伏学科最基础的知识内容延伸到最终的应用体系，涉及内容全面，涵盖面广，并通过课后阅读，进一步扩展读者的知识面。

（3）词汇丰富。本书所选词汇几乎涵盖了光伏基础知识到应用的绝大多数专业英语词汇，并且单独列出了中英文对照的部分，因此既可以作为高等学校教材，还可以作为各行业读者的自学教材。

（4）具备知识扩展内容。本书通过课后练习题以及课后阅读等内容的添加，为读者进一步对所学章节进行思考、并进一步延伸扩展所学内容奠定了良好的基础，为高等学校学生、工程技术人员在学习之后思维延伸，了解科研前沿动态提供了条件。

由于作者水平有限，书中难免有不足和疏漏之处，恳请各位专家、同仁和广大读者批评指正。

编者

2018年6月

Contents

Chapter 8

Chapter 1

Introduction to Photovoltaics

In *National Energy Law*, the explanation of new energy is: New energy refers to unconventional energy whose exploration and utilization is based on new technology. Such energy includes as wind energy, solar energy, ocean energy, geothermal energy, biomass energy, hydrogen energy, nuclear fusion/fission energy and gas hydrates etc. The characteristic of new energy is less pollution, large reserves.

For solar energy, it includes photovoltaics and solar thermal. Photovoltaics comprises principally the technology that generates direct current electrical power from semiconductors when they are illuminated by photons. As long as light in the solar spectrum is shining on the solar cell, it generates electrical power. When that light stops or becomes too dull, the electricity stops. Solar cells never need to recharge like a battery does. Some have been in continuous outdoor operation on earth or in space for over 30 years.

Specialized Vocabulary:

- conventional energy 常规能源，传统能源
- unconventional energy 非常规能源
- new energy 新能源
- renewable energy 可再生能源
- nuclear energy 核能
- wind energy 风能
- solar energy 太阳能
- biomass energy 生物质能
- ocean energy 海洋能
- geothermal energy 地热能
- hydrogen energy 氢能
- nuclear fusion/fission energy 核聚变/裂变能
- gas hydrates 天然气水合物
- photovoltaics (PV) 光生伏打；太阳光电
- photovoltaic (PV) 光生伏打的；太阳光电的

1.1 Present situation and trend of PV

According to the prediction of European Union (EU), solar power will account for more than 10% of the total energy consumption while renewable energy will claim 30% of the total energy supply by 2030. In addition, by 2050 solar power will account for more than 20% of the total energy consumption, renewable energy will claim 50% of the total energy supply.

Figure 1-1 is diagram of all different energy types. It is obvious that PV plays a major role in energy mix. It starts to be significant around 2030, and it will become the dominant energy form near this end of the century. There are many factors that will influence this data but the general development trend is similar. Therefore, solar power will be applied in large scale commencing in about 2015 and especially after 2030, and it could be the dominant energy in all likelihood after 2050. However, the net cost of PV needs to be low enongh and comparable to that from the existing grid in order to keep energy affordable.

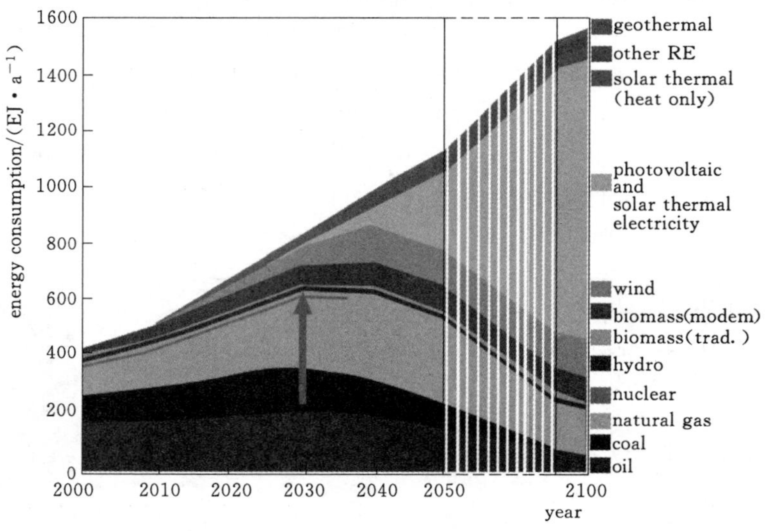

Figure 1-1 Scenario of future energy supply in the world. [1]

Specialized Vocabulary:

- PV power generation 光伏发电
- solar power 太阳能发电
- total energy consumption 总能耗、能源消费总量
- total energy structure 总能源结构
- geothermal 地热
- solar thermal 光热，太阳能热，太阳热

- dominant energy 主导能源
- grid 输电网
- State Grid 国家电网（公司）
- energy mix 能源组成
- hydro 水
- natural gas 天然气
- coal 煤
- oil，petroleum 油，石油

The use of solar energy is less than 1/1000 of the exploitable amount，therefore development and utilization of solar energy in China is worth expecting. For solar irradiation，most areas in China are medium to high irradiation areas except Guizhou plateau area and Chongqing.

China is located in the northern hemisphere and in generally，in the mid－latitude area，and the sun resource is rich with average annual solar radiation of about 5.9kJ/m^2. The solar radiation of valley areas in Tibet is as high as 7.5～7.9kJ/m^2，areas with the lowest sunshine such as Chongqing and Sichuan basin can also receive solar radiation of 3.3～4.2kJ/m^2（shown in Table 1－1）.

Table 1－1　Annual radiation of the solar energy resource belts in China.

resource belt	classification	annual radiation /($MJ \cdot m^{-2}$)	resource belt	classification	annual radiation /($MJ \cdot m^{-2}$)
Ⅰ	resource rich Ⅰ	≥6700	Ⅲ	resource rich Ⅲ	4200～5400
Ⅱ	resource rich Ⅱ	5400～6700	Ⅳ	resource starvation Ⅳ	<4200

The regional distribution of the quantity of China's solar energy radiation is：The solar radiation of the western region is highest and then comes that of the eastern part，while the the solar radiation in the northern region is still higher than that of the southern part，and the solar radiation in the plateau area is higher than that of the plains.

Specialized Vocabulary：

- solar energy distribution 太阳能分布
- solar energy resource(s) 太阳能资源
- northern hemisphere 北半球
- mid－latitude area 中纬度地区
- solar radiation 太阳辐射
- solar energy radiation quantity 太阳辐射量
- regional distribution 地域分布，区域分布
- plateau area 高原地区

1.2 The necessity and urgency of new energy development in China

1.2.1 Imbalance between supply and demand of energy in China

1. *Pressure of rapid growth on China's energy demand*

Since 2000, China's energy needs have maintained a rapid average growth rate of 9.4% annually, and it is expected that the total energy need will be doubled in 2030 on the basis of present usage. Compared with European countries, United States and other developed countries, China has greater economic development pressure and qrowth demond of the energy market.

Figure 1-2 and Figure 1-3 below are show the growing trend of energy demand in the world andprimary energy consumptions of different countries. It is obvious that energy consumption is growing faster than energy production, and the primary energy consumption in China is growing fastest.

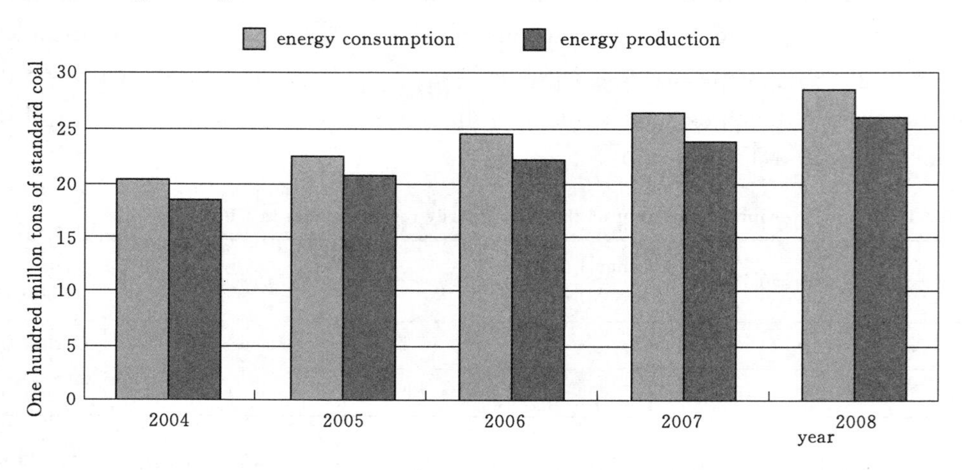

Figure 1-2 Growing trend of energy demand.
(Source: IEA World Energy Outlook 2010.)

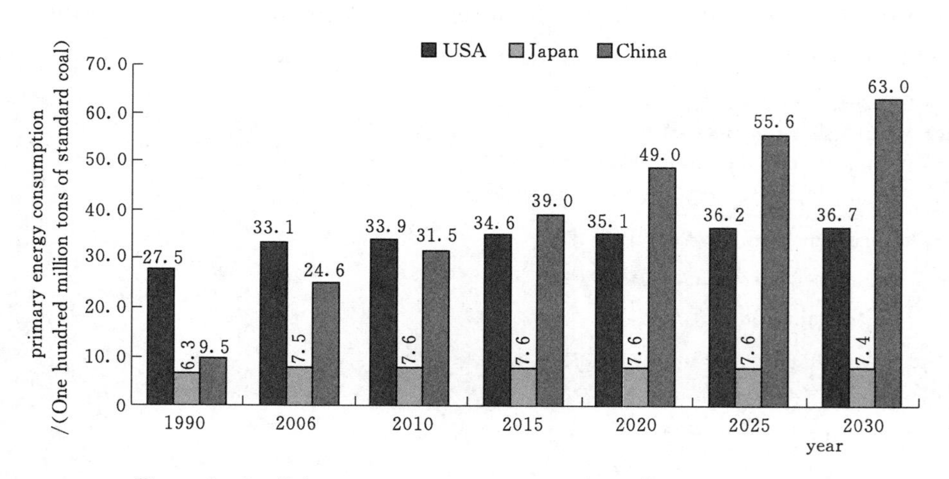

Figure 1-3 Primary energy consumptions by different countries.

The total energy demand scenario for primary energy in China and the world is shown in Figure 1 - 4. The compound annual growth rate from 2008 until 2035 shows that China's primary energy demand increases rapidly at least till after 2020, during which period there is also a significant growth in percentage of primary energy demand.

item	The total energy demand of primary energy/Mtoe							Compounded annual growth rate/%
	1990	2008	2015	2020	2025	2030	2035	2008—2035
global amounts	8779	12,271	13,776	14,556	15,263	16,014	16,748	1.2%
China	872	2131	2887	3159	3369	3568	3737	2.1%
China's percentage	9.9%	17.4%	21.0%	21.7%	22.1%	22.3%	22.3%	

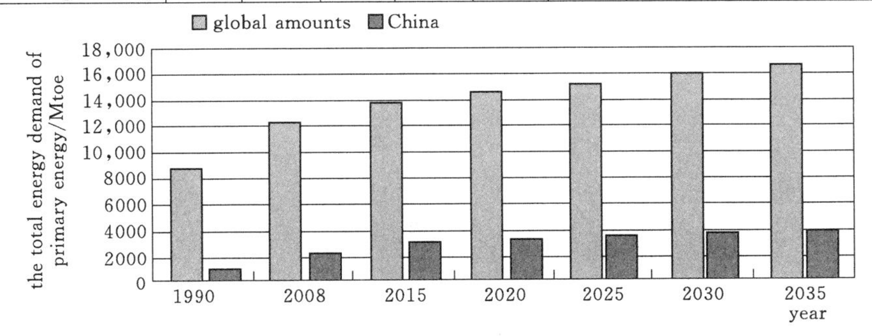

igure 1 - 4 The total energy demand scenario for primary energy in China and in the world.
(Source: IEA World Energy Outlook 2010 - New Policies Scenario, SEMI.)

The total electricity demand scenario in China and the world from 2008 to 2035 is shown in Figure 1 - 5. The annual compound growth rate of the global electricity demand amounts to 2.1%, and China's growth rate is much higher in the same period, with 3.8%. The absolute value of China's increases in power generation capacity is the highest in the world.

item	Total electricity demand/(TW · h)							Compounded annual growth rate/%
	1990	2008	2015	2020	2025	2030	2035	2008—2035
global amounts	11,281	20,183	24,513	27,733	30,016	32,696	35,336	2.1%
China	650	3495	5721	6949	7900	8776	9594	3.8%
China's percentage	5.5%	17.3%	23.3%	25.4%	26.3%	26.8%	27.2%	

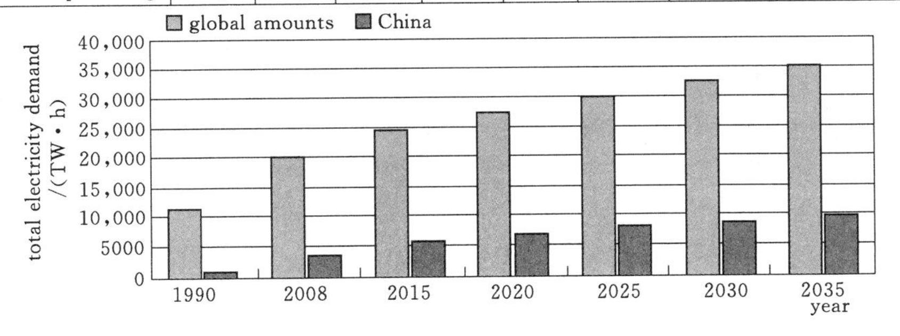

Figure 1 - 5 The total electricity demand scenario in China and in the world.
(Source: IEA World Energy Outlook 2010 - New Policies Scenario, SEMI.)

Specialized Vocabulary:

- energy demand 能源需求
- rapid growth 快速增长
- growing trend 增长趋势，发展趋势
- annual compound growth rate 年均复合增长率
- the absolute value 绝对值
- power generation capacity 发电容量
- total electricity demand 电力需求总量
- watt hour 瓦时

2. *Inadequate development of renewable energy in China*

Figure 1 – 6 shows the total installed capacity and new power equipment capacity of China in 2010. From the pie chart, it is obvious that the traditional thermal power occupied a large proportion of both China's total installed capacity and newly – installed power generation equipment, while the development of new energy (non – fossil energy) has been inadequate.

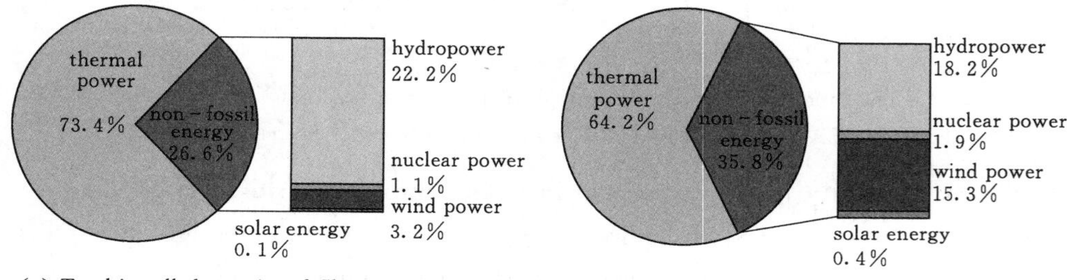

(a) Total installed capacity of China in 2010 (b) New power equipment capacity of China in 2010

Figure 1 – 6 The total installed capacity and new power equipment capacity of China in 2010. (Sources: National Energy Administration.)

The situation in Europe is different from China. In Figure 1 – 7, the proportions of Europe's traditional thermal power (such as coal and natural gas) and new energy (PV, wind power, Hydropower, nuclear power, etc.) are comparatively equal in the aspects of total installed capacity of power – generation and new capacity of power generation equipment.

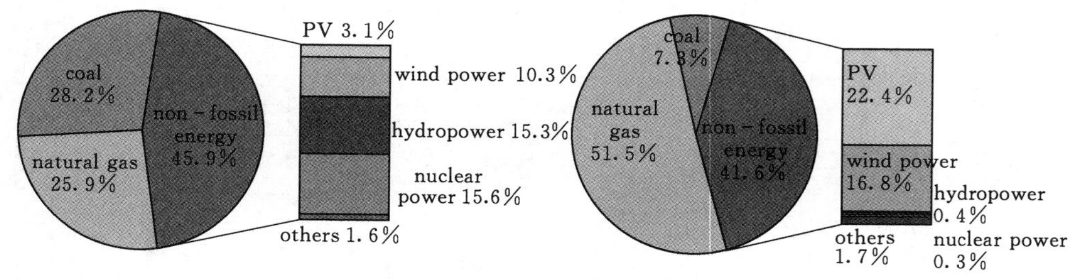

(a) Total installed capacity of Europe in 2010 (b) New power equipment capacity of Europe in 2010

Figure 1 – 7 The total installed capacity and new power equipment capacity of Europe in 2010. (Sources: EWEA, SEMI.)

Specialized Vocabulary:

- thermal power 火力发电
- total installed capacity 总装机容量
- newly-installed power generation equipment 新增发电设备
- fossil energy 化石能源
- non-fossil energy 非化石能源
- hydropower 水电
- nuclear power 核电
- wind power 风电
- natural gas 天然气

3. *Serious shortage in China's fossil energy supply*

Table 1-2 shows the reserves, production and production life of fossil energy in China and the world. From the data shown in Table 1-2, the production life of China's fossil energy is as short as 10~40 years, far shorter than that of the world average level.

Table 1-2 The reserves, production and production life of fossil energy.

fossil energy	reserves	production	production life/y
global coal/10^8 t	8260	67.705	122
China's coal/10^8 t	1145	27.927	41
global crude oil/10^8 t	1708	40.667	42
China's crude oil/10^8 t	21	1.897	11.1
global natural gas/10^9 m^3	185.02	3.063	60.4
China natural gas/10^9 m^3	2.46	0.07	32.3

In all the above cases, the solution to the imbalance between energy supply and demand of energy in China is to develop new energy in large scale.

Specialized Vocabulary:

- reserves 储量，探储量
- production 生产量
- production life 开采年限
- crude oil 原油
- shortage 短缺

1.2.2 The need to cope with climate change

Due to the wide use of fossil fuel that products greenhouse gases for energy production, the situation of coping with global climatic change is very serious, and green new energy is urgently needed. Therefore, *Kyoto Protocol* proposed emission targets in

1997. Initially, each contracting party was asked to reduce greenhouse gas within four years (from 2008 to 2012) by at least 5% compared with the level in of 1990. Secondly, there were different standards for each contracting party. Thirdly, it proposed three International flexible mechanisms, namely the clean development mechanism, joint performance mechanism, and international emissions trading.

Subsequently, the Copenhagen Conference proposed emission targets in 2009. It included setting up emission reduction targets for advanced countries after 2012. The developing countries also agree to take substantive domestic measures with technological and financial support from advanced countries.

Figure 1 - 8 is the comparison in carbon dioxide emission of different energy resources. It is obvious that carbon dioxide emission from renewable energy is far lower than that of fossil fuel energy. China's carbon dioxide emission was 7.22 billion in 2010 and 80.5% was from coal burning. Therefore, large - scale development of renewable energy plays a vital role in reducing greenhouse gas emissions.

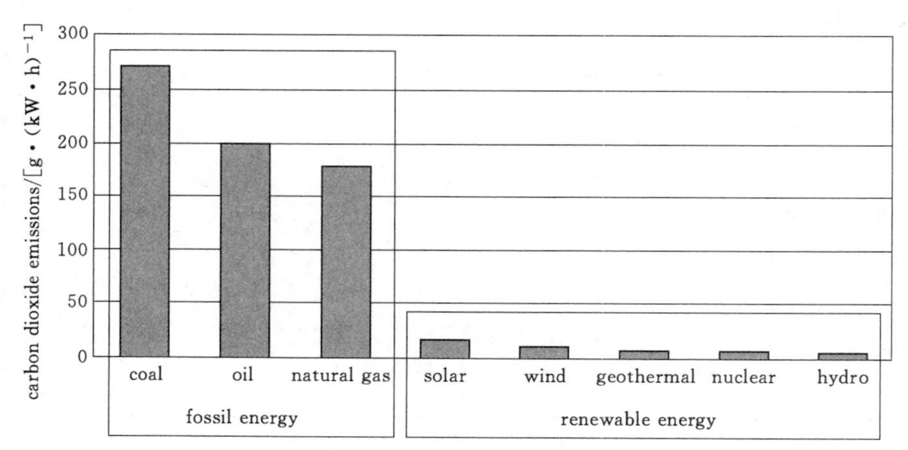

Figure 1 - 8 Carbon dioxide emission from different energy resources.

Specialized Vocabulary:

- greenhouse gas 温室气体（尤指二氧化碳或甲烷）
- emission reduction 减排
- carbon dioxide 二氧化碳

1.2.3 Urgent need to cultivate strategic emerging industries

The 1st industrial revolution, the steam engine, occured during 1760s, the 2nd industrial revolution relating to electric power was during 1870s, and the 3rd industrial revolution about information, commenced during 1940s. Now, there comes the 4th industrial revolution characterized by new and renewable energy, which will offer China an excellent opportunity to advance in world prominence.

Former China's Premier Wen Jiabao delivered a speech titled *Make Science and Technology the Leading Force of Sustainable Development* in November 2009, when new energy was given top priority among the seven strategic emerging industries.

1.3 Photovoltaic (PV) history

In 1839, French physicist A. E. Henri Becquerel put two pieces of metal into an ionic solution and made a voltaic (electric) battery which would produce extra voltaic potential when exposed to light (PV effect). In 1873, British scientist Wilough B. Smith noticed the light sensitivity of selenium (Se) material, and deduced that the conductive ability of selenium increases proportionally with the amount of light it receives when the selenium material is exposed to light. In 1883, Charles Fritts developed the first functional, intentionally constructed photovoltaic cells on the basis of selenium [2]. He melted Se into a thin sheet on a metal substrate and pressed an Au-leaf film on as the top contact. It was nearly $30cm^2$ in area. He noted, "the current, if not wanted immediately, can be either stored where produced in storage batteries, or transmitted a distance and used there". Later, people call those devices that have this effect "photovoltaic devices". Among them, the semiconductor p-n junction devices have the highest solar-power conversion efficiency, and these devices are often called "solar cells".

In the early 1950s, the researchers at Bell Laboratory of United States accidentally found that silicon is sensitive to light after the impurity processing when they were looking for reliable power supply for a remote communication system, and they discovered that the p-n junction diodes could generate a stable voltage. Within a year, they had produced a monocrystalline silicon p-n junction solar cell with the efficiency of 6%. From 1961 to 1971, the research was focused on improving radiation-resistance ability and cost reduction. From 1972 to 1976, all kinds of monocrystalline silicon photovoltaic cells for space use were developed. Ultra-thin monocrystalline silicon photovoltaic cells were developed in mid 1970s. In 1976, the Nobel Prize for physics was awarded to Professor Mott for discovering the electronic process of amorphous solidification. Then amorphous silicon photovoltaic cells came into being based on monocrystalline silicon and polycrystalline silicon.

Specialized Vocabulary:

- monocrystalline silicon 单晶硅
- polycrystalline silicon 多晶硅
- amorphous silicon 非晶硅
- photovoltaic cells 光伏电池
- photovoltaic devices 光伏器件
- solar-power conversion efficiency 太阳能转换效率；光电转换效率

- radiation - resistance ability 抗辐射能力
- ultra - thin 超薄
- voltaic potential 伏打电势
- voltaic battery 伏打电池
- solar cell 太阳电池

1.4 Solar energy utilization patterns

1.4.1 Energy conversion mode classification

1. *Solar energy light - thermal conversion*

Solar thermal devices receive and concentrated, then convert it into heat, which is further used for solar water heater, heating, refrigeration; solar drying in agricultural and sideline products, medicinal materials and wood; distillation, greenhouses, cooking; solar desalination; solar thermal power generation, etc.

2. *Solar energy light - electric conversion*

This is direct light - electric conversion (photovoltaic effects) and light - heat - electric conversion.

3. *Solar energy light - chemical Conversion*

Light - chemical conversion uses solar energy for decomposition typically of water to hydrogen with a catalyst, but it is still in research and experiment stage.

4. *Light biological utilization*

Light energy converts into biomass energy, fast - growing plants, oil crops.

Specialized Vocabulary:

- Solar water heater 太阳能热水器
- solar drying/solar energy drying 太阳能干燥
- solar desalination 太阳能海水淡化
- solar thermal power generation 太阳能热发电
- photovoltaic effects 光伏效应

1.4.2 Applications

1. *Space application*

The American satellite Vanguard - I was launched into orbit in 1958 with the solar cells as its power system for signaling, which means that new era of using solar cell for space power supply was opened up.

(1) Hubble Telescope. It has two solar panels on each side. The panels are 11.8m

long, 2.3m wide and produce 2.4kW power (shown in Figure 1-9).

Figure 1-9 Hubble Telescope with solar cell arrays.

(2) China's manned spacecraft - Shenzhou Ⅶ. Recently, China has launched Shenzhou Ⅶ at Jiuquan Satellite Launch Center in October 2016, the main power supply is from GaAs solar panels.

In Figure 1-10, the major power supply for Shenzhou Ⅶ is from its solar cell arrays. The two wings are composed of the panel components, 4 sections of solar panels for each wing, respectively. Solar arrays consist of a battery of monocrystalline silicon boron back surface field. The actual solar-power conversion efficiency is as high as 15%. The whole spacecraft use 11,690 interconnected solar cells, cloth coefficient reaches 92.5%. Its solar-power conversion efficiency and cloth coefficient have met the international advanced level for similar products.

Figure 1-10 China's Shenzhou Ⅶ manned spacecraft with solar cell arrays.

Subsequently, China has launched shenzhou number eleven at Jiuquan Satellite Lannch Center in October 2016, for which the main power supply is from GaAs solar panels.

Specialized Vocabulary:

- space application 空间应用
- space power 空间电源
- grid – connected generation 并网发电
- solar panels 太阳能电池板
- efficient monomer batteries 高效单体电池
- monocrystalline silicon boron back surface field 单晶硅硼背场
- solar – power conversion efficiency 光电转换效率，太阳能转换效率

2. *Ground application*

Ground applications including lighting, communications, transportation, Building Integrated/Attached Photovoltaic, and Grid – connected (solar farms) are the trends for PV technology application.

(1) Solar Lighting. Solar lighting takes solar energy as its power, as long as the sunshine is sufficient, the cells can be installed locally with a rechargeable battery, without involvement of the power supply grid, with no requirement to dig ditches or bury lines and no conventional power consumption, it is a kind of green environmental protection product.

1) Solar street lighting. Figure 1 – 11 shows a solar street light. It is comprised of solar modules, battery, charge controller, lighting circuit, poles, and other sensor and control components.

Figure 1 – 11 Solar street lights.

The pole is the supporting part of the whole system, which is different from the conventional street light, it needs to support not only the light holder, but also the solar modules and batteries, those are often near the base. Therefore, solar street lights at present preferably use the efficient crystalline silicon solar cells, as well as a choice of lamps of low powerconsumption and high brightness (e. q. LED) . The controller of solar street lights should be able to be switched on and off automatically in addition to being equipped with such functions as to prevent reverse charging, overcharge and over discharge, short circuit and reverse lamp which are possessed by common photovoltaic systems.

2) Solar garden light (shown in Figure 1 - 12) . The parameters of solar garden light are in below:

a. Solar modules: 2W.

b. Battery: preferably Ni/Cd but Li - ion and Pb - ion are also used.

c. Light source: LED.

d. Lamp body material: Al alloy.

e. Service time: >8h.

f. Continuous rainy days: 2~3 days reserved.

g. The lamp body height: ~0.6m.

3) Solar traffic warning light. The solar traffic warning light (shown in Figure 1 - 13) can confront wind, hail, ultraviolet irradiation, and must be unaffected by the environment, and sustain light for more than 10 days even over consecutive raining days.

Figure 1 - 12 Solar garden light.

Figure 1 - 13 Solar traffic warning lights.

(2) Solar bag, solar computer bag. The built - in multifunctional solar emergency charger (Figure 1 - 14), may supply power by converting solar energy into electricity The solar panels absorb solar energy at any time and anywhere, and the power can be stored in a built - in battery. This product is suitable for emergency use of power when you are in need of charging of cell phone, digital camera and other digital products outdoors.

The components of a typical solar bag are described as solar panels (waterproof), storage battery, charging adapter, connecting wiring, USB charging circuit, adapter, computer accessories (configuration of these accessories can be according to the customer specification) .

(3) Solar car and yacht.

1) Solar electric car. The solar electric car described herein is from the Institute of

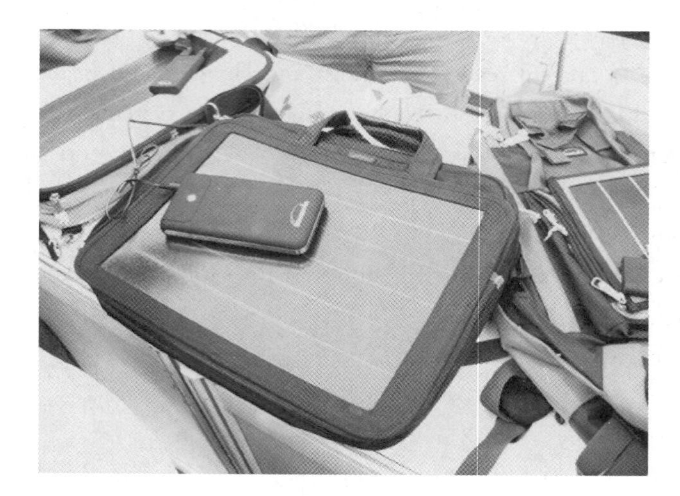

Figure 1 - 14 Solar bags.

Solar Energy Systems, Sun Yat - sen University (though there are now many demonstration solar cars) . Its appearance is like the battery car in the park, capable of holding six passengers, but the highest speed is only 48km/h, with travel time as short as one hour, shown in Figure 1 - 15. Of course, there are many recent advances.

Figure 1 - 15 Solar electric cars.

2) Solar yacht. The first solar yacht made in China came into being in Zhuhai in July 2008. It was also one of the first solar yachts launched into market in the world, before this most of the foreign solar yachts and boats were used for scientific experiment purposes or tourism promotion. Although in aspect of speed, the solar yacht may not yet satisfactory level, the solar yacht made itself the star at the international boat show in Miami due to its "green" and environmental protection characteristics.

The yacht's sails are the key parts whose purpose is to concentrate direct and indirect sunlight, and they can also be used as a sail. The array is controlled by computer, it can rotate around two axes, and the angle can be altered for better light concentration. The energy collected by solar modules is stored in lead - acid batteries. Solar yacht is shown in Figure 1 - 16.

Figure 1 - 16 Solar yacht.

(4) BIPV, BAPV. Solar cells become part of the building itself in the application of Building Attached Photovoltaics (BAPV), and Building Integrated Photovoltaics (BIPV), such as roof tiles, curtain walls, doors, windows, and awning etc. There are many examples for BIPV and BAPV including shown below in Figure 1 - 17.

(a) Yingli PV research and development center

(b) Awnings: Cover sunshine, keep off rain

(c) Photovoltaic curtain wall

(d) PV corridor

Figure 1 - 17 Four examples for BIPV and BAPV.

Specialized Vocabulary:

- ground application 地面应用
- solar light 太阳能灯
- solar street light 太阳能路灯
- solar modules 太阳电池组件
- battery 蓄电池
- charge controller 充放电控制器
- lighting circuit 照明电路
- poles 灯杆
- efficient crystalline silicon solar cells 高效晶体硅太阳电池
- low power - consumption 低功耗
- high brightness 高亮度
- lighting lamps 照明灯管
- reverse charging 反充
- overcharge 过充
- overdischarge 过放
- switch on/off 开/关
- solar garden light 太阳能草坪灯
- solar traffic warning lights 太阳能交通警示灯
- ultraviolet irradiation 紫外线照射
- sustainable light 可持续发光
- built - in 内置
- multifunctional solar emergency charger 多功能太阳能应急充电器
- emergency charger 应急充电器
- convert solar energy into electricity 太阳能转换成电能
- absorb 吸收
- built - in battery 内置蓄电池
- battery (powered) car 电瓶车
- solar electric car 太阳能电动车
- solar yacht 太阳能游艇
- launch into market 投放市场，投入市场
- environmental protection 环境保护
- sail 帆
- lead - acid battery 铅酸蓄电池
- Building Attached Photovoltaics (BAPV) 光伏与建筑结合/一体化
- Building Integrated Photovoltaics (BIPV) 光伏建筑一体化
- photovoltaic curtain wall 光伏幕墙
- PV corridor 光伏走廊

(5) The application of photovoltaic power generation in Shanghai World Expo. Solar power generation projects of Shanghai World Expo are examples of BIPV of the largest scale and highest technology in China. Solar power generation technology was applied more widely at the Shanghai World Expo than in the history of all the world expos. Chinese Pavilion with PV technology is shown in Figure 1 - 18.

Figure 1 - 18 Chinese Pavilion with PV technology.

The solar power generation technology was applied to the whole World Expo park, World Expo center, Chinese Pavilion and Theme Pavilions. Among all these, justin the expo center, there was a large solar project on its roof with capacity of 1MW, comprising a large area of efficient solar modules are laid on its roof, and on its walls were installed double - faced transparent sunshade solar modules. The array covers an area of $8000m^2$ and is able to generate electricity of 1 million kW・h, and cut carbon dioxide emissions annually by 900 Theme Pavilions with PV technology is shown in Figure 1 - 19.

Figure 1 - 19 Theme Pavilions with PV technology.

A "seven - colored flower" is showed in Expo Jiangsu Pavilion in 2010 Shanghai World Expo, demonstrated to people how solar energy is captured across the light spectrum, and the seven colors of red, orange, yellow, green, indigo, blue, and purple are reflected. This flower is in fact a solar device in the form of sunflower that is made of solar panels. It can rotate around the sun, collecting seven colors of spectrums to gather new energy. Solar energy absorbed is transmitted to the daylighting import & storage system, where it can be converted into power (shown in Figure 1 - 20) .

Figure 1 - 20 Jiangsu Pavilion with PV technology.

The Japanese Pavilion, was a building known as "Purple silk - worm island", the outside of it is made of semi - transparent films that can absorb solar power and shine at night. The curved dome has three holes at three different angles. The holes are to receive rain for recycling use, and introduce sunlight for illumination to reduce electricity demand. The angled holes are to enhance the circulation of warm and cold air and to reduce air conditioning energy consumption (Figure 1 - 21) .

Figure 1 - 21 Japanese Pavilion with PV technology.

Specialized Vocabulary:

- solar (power generation) projects 太阳能发电项目
- capacity 容量
- efficient solar modules 高效太阳电池组件
- solar devices 太阳能装置
- solar panels 太阳电池板
- daylighting import system 采光导入系统
- convert into power 转化为电力
- energy consumption 能耗
- generate electricity 发电
- cut carbon dioxide emissions 二氧化碳减排

(6) Photovoltaic power station (solar farms). Areas with no electric power in the west of China will depend to a large extent on photovoltaic power station to provide electricity.

By 2012, the total photovoltaic power station capacity of Qinghai had grown to over 500 million kW·h, with the installed capacity of photovoltaic power station being 7% of the provincial total.

At present, there are 42 grid-connected photovoltaic power stations in Qaidam basin in Qinghai province, with capacity of 1003MW. Qinghai has made several achievements as follows, creating a photovoltaic power station with largest installed capacity within the region, creating the largest grid connected photovoltaic power station in the world at the date of installation, the first grid-connected photovoltaic power station of mega watt capacity in the world. PV power station in Qinghai Province is shown in Figure 1-22.

Figure 1-22 PV power station in Qinghai Province.

Specialized Vocabulary:

- photovoltaic power station 光伏电站
- grid - connected 并网（发电）
- installed capacity 装机/容量
- solar energy utilization patterns 太阳能利用方式
- the total capacity 总发电量

1.5 Reading material (Please read the article and find out the specialized vocabulary)

1.5.1 Basic understanding of photovoltaic generation (From EECA energywise, the power to choose)

Solar electricity generation (photovoltaics)

How PV works

PV cells convert sunlight into electricity by an energy conversion process. In most PV cells, photons (light energy) hit the cells, exciting electrons in the atoms of a semi - conducting material. Silicon is the most commonly used semi - conductor. The energised electrons result in the generation of an electrical voltage. In other words, electrons flow, producing direct current (DC) electricity.

PV cells are usually fairly small, with lots joined together to form a PV panel. These panels are then grouped together into PV arrays.

Where PV is used

PV panels and arrays are used in both stand - alone power systems and grid - connected generation systems, as well as in other small applications, like weather stations, some road signs and parking meters.

Key components of PV systems

PV systems are typically made of these key elements:

- PV panels, cables, and mounting or fixing hardware
- An inverter and controller
- Batteries, back - up generators, and other components in off - grid situations
- Special electricity meters, in the case of grid - connected systems

Capacity rating of PV panels

Each PV panel (or module) is rated on its peak electrical output under standard test conditions. For example, a module with a 75 Watt peak rating (75Wp) will have an output of 75 watts under standard test conditions. Modules are available in sizes from 5Wp to 200Wp.

A typical domestic system is around 1000Wp to 3000Wp. To get a total combined capacity of 1000Wp you will need to buy a number of smaller modules and connect them to form a "solar array" .

How much electricity do PV panels produce?

The amount of electricity a PV system generates depends on the intensity of the sunlight to which it's exposed. PV still produces electricity on cloudy days, but less than with direct sunlight.

Obviously PV panels generate no electricity at night, and less in the morning and evening than in the middle of the day. How much electricity you will get from your PV panels per day can be worked out when designing your system.

In a whole day, a well - located PV panel will typically generate between 2. 5 and 5 times its rated power output. So a 1kWp (kilowatt peak) PV panel could produce between 2. 5kW • h (kilowatt hours) and 5kW • h per day, or between 880kW • h and 1750kW • h per year.

Where can I use PV panels?

PV panels work well in both rural and urban conditions. The best places to use PV is in places that get a lot of sunshine each year and where the sky is generally clear rather than cloudy.

PV works best in north - facing places with year - round sun. Panels are usually installed on roofs but can also be placed on facades, conservatory roofs, sun shades, garages or specially - built stands on the ground.

Make sure your site:

- Faces north (south - facing panels are for the northern hemisphere)
- Is free from shade and exposed to good sun all year
- Has enough space - a typical 1kW unit needs an area of around eight square metres.

If you are designing a stand alone power system, you will probably need to combine your PV array with other generators - such as a small wind turbine, micro - hydro system, or a petrol, diesel or biodiesel generator.

Costs

- The cost of solar panels has fallen significantly in recent years as production expands and technologies improve. However installation and related equipment, such as a battery bank (if required), add to the cost considerably.

1. 5. 2 Basic understanding of photovoltaic generation (From Internet)

General situation about solar energy in China

China's status as a world power is undeniable. As the whole issue of energy crisis acquired a demonic proportion with each passing day, all eyes turn towards the gigantic manufacturing nations, and China attracts a lot of attention in all such discussions.

China lies in the northeastern part of East Asia between 48 and 538 North latitude and 738 to 1358 East longitude with an area of 9. 6 million km^2. According to the data of Chinese weather bureau (CWB), the total solar energy resources are enormous in large soil area, but the irradiation is various in different zones. The Tibet and southeast of the Qing-zang altiplano lie in the highest irradiation zone of solar energy, and the annual hours of sunlight is more than 3200, and the annual irradiation amount is about 6600～8500MJ/m^2. The annual hours of better irradiation zone are about 3000～3200, and the annual irradiation amount is about 5800～6600MJ/m^2. The available zone is about 2200～3000, and the annual irradiation amount is about 5000～5800MJ/m^2. The deficient zone has a share of 33%, and the sunlight hours are less than 2200, and the irradiation amount is less than 5000MJ/m^2. However, according to the data of Chinese development and innovation committee in 2006, the abundant zone of solar energy has a share more than 67%, which is the comparative efficient zone of solar energy application in China, and the sunlight hours is more than two thousands, and the annual total amount of irradiation is more than 6 billion MJ/m^2, so China has abundant solar energy. Certainly, China has thousands of towns and hundreds of cities and the different cities have the different daily irradiations and best obliquities. According to the different latitudes, Chinese main cities have different solar irradiation parameters.

Although China has extremely rich solar energy resources, China's new electricity generation capacity is still coming predominately from fossil fuels. As a result, renewable electricity capacity and generation considered as a share of total capacity and generation decreased instead of increasing. With this fact in view, we have a reason for not being optimistic about China's carbon emission future. In 2005, coal made up about 68. 7% of China's total primary commercial energy consumption, while in the other countries in the same year it was only about 21%. Coal has the highest carbon intensity among fossil fuels, resulting in coal - fired plants having the highest output rate of CO_2 per kWh. This situation creates a serious threat to global warming and is a very important case because the effects of global warming are clear.

Therefore, China's Energy Ministry is in pursuit of a solution to the imminent energy crisis. The recent policies regarding simplification of approval grants to solar power projects seem to be steps in the right direction. On similar lines, the process of setting up power plants based on conventional means such as fossil fuels has become difficult. This is an obvious attempt to divert people and organizations towards solar energy and other environment friendly energy forms through the provision of economic enablers in the form of subsidies and incentives.

Thousands of years ago, the solar energy is used to insolate the corn and salt and clothing by human. One thousand year ago, the coppery concave mirror is used to obtain fire by Chinese ancestor, which is collected from spring and autumn to Song dynasty by Chinese museum. But the easy use of solar energy in China is not change until 1971, and the first application of PV is utilized to the power supply of secondary planet by Chinese scientist. The PV is first utilized to the ground in 1973. By the past nearly 40 years, there are many applications for the direct and indirect utilization of solar energy, and the application domain of solar energy is increasing rapidly with the development of China, Such as solar water heater, water pumping, road lighting system, solar heating buildings, solar refrigeration, air conditioners and PV generation system. In China, mostly the solar energy is used by the solar water heater and solar energy greenhouse. The extensive utilizations of solar energy have brought great environmental and economic benefits in the recent decades.

The utilizations of solar energy can be divided into two kinds. One is the application of solar energy heat, which can be divided into the direct and indirect utilization of solar energy heat. Such as solar energy hearth, solar energy house and solar energy greenhouse, there are the direct utilization of solar energy heat. The indirect utilization in China includes some domains, such as solar energy desiccation, solar energy calefaction of industry, solar energy refrigeration of industry and solar energy heat generate electricity by using solar energy collect heater. The other one is PV generation electricity, which is used to generate electricity by solar cell. Certainly, there are many utilizations in China, such as space domain, navigate assistant of sea, wireless communications, portable power supply, cathode protection, PV water pump and lighting. At present, China's PV industry is growing faster than perhaps any other country in the world. So, there are many encouraging signs, as well as many critical challenges, for both the international and indigenous photovoltaic industries in the energy markets in China.

The PV industry of China has a huge development in past 10 years. For example, the yield of Chinese PV in 2007 is more than 1200MW, and which has share of 35% in whole world, which ranks the first in the world. The government encourages the development of new and renewable energy in the built environment. Also, The UNDP (United Nation Development programme) supports the Chinese government in its obligations in the field of environment and energy. It focuses on the promotion of sustainable energy for sustainable development, for example, the promotion of renewable energy and energy efficiency. The market share of Chinese PV has increased from 1% to 35% in the last 8 years, and the quality has step up at the same time. Programs like "Golden Sun" also work towards popularizing the concept of solar energy. Granting subsidies to people who install solar power systems at their homes is also a step in the right direction, and has already given favorable results. The offsetting of installation costs associated with large scale solar power plants through increased responsibility on other well settled sectors like construction and finance also promises to rope in more and more households into the list of solar powered ones.

PV power generation will play a significant role in China's future energy supply. According to the present plan, total PV power installations will reach 1.8GWp by 2020 and 1000GWp by 2050. According to forecasts made by the Chinese Electric Power Research

Institute, renewable energy installations will account for 30% of total electric power capacity in China by 2050, of which PV installations will account for 5%. At present, the biggest photovoltaic plant is established in Shilin of Yunnan province. The capacity and the investment are 66 MW and 0.6 billion dollars, respectively. Consequently, the market and development potential of solar energy are startling in the future China.

1.5.2.1 PV utilization in China

The pioneer application of PV in China is utilized to the exploitation of space domain by Chinese scientist due to the price of PV is very costly, such as which is used to the power supply of secondary planet and spacecraft, and the weight of PV has a share of 10%~20% of the whole secondary planet weight, and the price of PV has a share of 10%~30% of the whole secondary planet price. So the weight and price of PV must be decreased in order to decrease the launch price of secondary planet, which is the research direction of space application in future. PV has been used in various domains with the development of economy and society at present.

1. *The city road lighting system*

It is well known that hundreds of big cities lie in the large soil of China, and there are more than 10 million street lamps in those cities. The annual total amount of whole city road lighting system is more than one billion degrees, and billions of dollars is expended in order to pay the expense of electric power, which had became the enormous burden of the Chinese government. The abundant primary resource is consumed at the time, and the wastage of coal is more than two milliontons, and a mass of CO_2 and SO_2 and NO_x are vented, and a-mass of castoff pollute the environment and water and air. Fortunately, the problems have been considered by the Chinese central government and local government and ordinary people. The solar energy is utilized in the city road lighting system by some local governments in order to improve the local environment, i.e., solar energy street lamp, solar energy community lighting and solar energy scenery lighting. The solar energy street lamp has better competition and is more popular. More and more cities in China begin to replace the conventional street lamp by using the solar energy street lamp. For instance, there are more than 3000 solar street lamp by using the city lighting system in Binzhou. The whole street lighting system is replaced by using the solar street lamp in Linan, Zhejiang Province. Moreover, the annual electric power cost of conventional street lamp in Hangzhou is 0.3 billion RMB. It is estimated that the investment of solar LED street lamp is equal to the conventional lighting system during 3 years. The great economy income is received from the renewable solar street lamp during the remaining years. Synchronously, the enormous income of environment is gained.

Other solar energy lighting systems have been used to improve the life of common people, such as court lighting, lawn lighting and scenery lighting. The solar lighting systems improve the habitation and life quality of citizen, which brings the enormous environment and economy benefit. For example, the solar street lamp is used to improve the lighting condition of Shitai freeway in 2006, and the total investment is more than three million RMB. The total of traffic flux is increasing 13% compared with the corresponding period of last year, and the annual total of traffic accident in winter is decreasing from 50 to 25, and the economy loss is decreasing from 2 million RMB to 12,000RMB, and the total of injured people is decreasing from 41 to 25 at the same time. Moreover, the scenery lighting is used to improve the sight of hilly country park in Xiamen, and the total of solar scenery lamp is more than 200. The lighting system of remote village in Yangzhong is achieved by using the solar street lamp. As mentioned above, the requirement of solar street lamp is enormous with the sustainable development of China in future.

2. *Solar water pump*

West zone of China include Tibet and Xinjiang and Qinghai and Shaanxi and Gansu and Sichuan, which is the best underdeveloped zone in the whole China, the economy and zoology is very brittle. And the area of West zone is more than 4 million km^2, and the population is more than 0.2 billion, and the zone contains abundant natural resource. Unluckily, the water is exceeding lack in the northwest zone of China, and the desert zone of northwest China in 2007 is more than 1.3 million km^2. Based on the data of Chinese forest bureau, the annual increscent desert area in 2001 is more than 2000km^2. For instance, Xinjiang lies in the northwestward of China, the area is more than 1.6 million km^2, and the natural resource is abundant. But the Take Lamagan desert lies in the middle part of Xinjiang, and the area is more than 0.33 million km^2, and Xinjiang divides into 2 parts. Fortunately, Chinese central government and local government have realized the problem of increasing desert area, and some actual actions have been implemented to improve the environment and zoology of northwest zone. And the groundwater of desert zone is abundant in Chinese northwest zone. The groundwater gives a hope to banish the desert. Some electric power is used to pump groundwater by using water pump. As mentioned above, the environment and zoology is very brittle, and the desert zone is far away from main power lines, and the traffic is not convenient.

Simultaneously, the high solar irradiation exists in the large northwest zone. The solar water pumping has great potential to banish the desert and improve the irrigation area of northwest farmland. Some actual applications have been used to improve the zoology of northwest and traffic, such as the desert road, protect desert oasis and solidify desert. For instance, the Talimu desert road is laid to link the north and the south of Xinjiang in 1995, and the length of desert road is more than 550km. At present, more than 5 desert roads have been laid in the last 20 years.

In an example of solar water pump application in remote villages of northwest China. The area of grassland in China is more than 0.4 billion hectares. The area is 0.102 billion hectares, which is irrigated by local people. However, the environment and zoology of Chinese northwest are brittle due to the annual rainfall is less than evaporation. The area of desert is increasing in the last 50 years, and the total of desert area in 2007 is more than 1.3 million km^2.

The water pump for irrigation is used to improve the local environment by local people. It is well known that the economy of Chinese northwest is very poor, and the local people cannot support the costly fee of diesel water pump, and the diesel oil will destroy the brittle zoology of northwest. In other words, solar water pump gives a hope to the people to improve the local zoology. But the solar water pump has a higher price than the diesel water pump. Fortunately, the central government and local government give some assistance to increase the popularization of solar water pump. An actual solar water pump system is described in the section, which lies in Neimenggu Province. The end user' name is Nashun, and the grassland area of family is 5.3ha. The system contains: a well, a PV, a water pump, a converter/inverter and some sprinklers. The capacities of PV and water pump are 3 kW and 500W/24V, respectively. The total of investment in 1999 is more than 60, 000 dollars. The total of livestock is more than 370 by using the solar water pump. In China, the irrigation area of solar water pump in 2003 is 534ha. The object area in 2010 is more than 392,000ha, and the need of PV is more than 261MW. As mentioned above, the prospect of solar water pump in China is great in future.

3. *Distributed generation* (*DG*)

The large - scale distribution network (LSDN) is considered by Chinese government in past 30 years, and the accumulative total amount of electric energy in 2007 is more than 0.713 billion kW. However, a potential danger exists in the LSDN because the modern people are more and more dependent on the electric power supply. If an electric network occurs an accident, which will affect the daily life of millions of people, and the unpredictable accident will stop the factory production and the society movement because the electric power is cut. For an instance, the northeast of USA and the east of Canada are cut the electric power by an unpredictable electric network accident in 2003 and more than 50 million people are affected during the power cut, and the daily economy loss is more than 30 billion dollars. So a credible power supply must be found in order to conquer the unforeseen accident. Fortunately, the solar energy is not big affected in the natural disaster and accident. And a solar distributed generation can partially afford the electric supply. With the improvement of people life, more and more people and Chinese government have realized the important of DG to improve the security of electric power supply. For instance, millions of cattle farmers working in the widest northwest zone of China, the herd and cattle farmer will move with various seasons. Because the browse zone is far away from main power lines, so they can conveniently gain the electric power by using the small DG units. In a word, the DG is important to improve the security of electric power supply and the life quality of common people.

Some actual applications have improved the life of ordinary people, who located in remote villages of Chinese northwest zone, such as mobile vehicle of power supply, region power supply and no watch transformer substation. For instance, the DG has been used to the national defence of China. It is well known that China has more than 5000 islands, which intersperse among the 3 million km^2, and mostly the island is garrisoned by the People Liberation army (PLA) . Thousands of PLA garrison the island in order to safeguard the coastal areas and territorial seas. But the life condition of PLA is very hardy due to the area of most of the islands are very small, and where they have not fresh water and fossil resources. Fortunately, the small islands have abundant solar resource. The DG is the best way to improve the life quality of PLA and the islander. The PV DG has been used in thousands of island army. Some other actual applications have improved the life of soldier and islander.

At present, the seawater is desalted in order to provide enough drinking water, and the electric power drive thousands of martial equipment, such as radar, computer and missilery. Certainly, the standard of living is increasing by using the DG. In an example of no watch transformer substation and railway station in Tibet of China. In 1 July 2006, the Qing-zang railway is established from Xining of Qinghai Province to Lhasa of Tibet by thousands of worker, and the length is 1956km, where we have execrable environment and far-flung winter. In a word, the solar DG has great potential in future China.

4. *Grid-connect PV generation (GPG)*

At present, the GPG is regarded by the developed country in the recent decades. The GPG has a biggish share in the whole yield of PV, and which will achieve a great development in future. However, the development of GPG in China is very slow, and the market share is only 0.3% in the last 30 years. The essential reason is the costly electrovalence of PV. At present, the electrovalence of PV is about 0.6 dollars per degree, and which is too high to support by the common people. Because of the electrovalence of the conventional fossil resource in China is only 0.5 RMB. Fortunately, Chinese central government had realized the problem, and some hortative policy is established, such as the generating electric power of PV must be accepted by Power Company, and the price is enhanced in order to ensure the advantage and enthusiasm of investors. The desert zone of northwest China in 2007 is more than 1.3 million km^2. The capacity total of PV is 100MW per km^2. If the fixed PV area of desert has a share of 1%, and the capacity total of PV is 13,000GW. In other words, the capacity is double compared with the accumulative total of electric power at present. With the improvement of technology and the decreasing price of PV generating electric power (PGEP), the prospect of large-scale desert PGEP is enormous in future China. At present, three PV power plants establishing in west desert, the capacity is more than 20MW. The object capacity of the desert PGEP is 200MW in 2020. For example, Yangbajin desert PGEP is established in Tibet, the capacity is 100kW. The architecture area in China is more than 40 billion m^2, and housetop area is more than 4 billion m^2, and the area of southerly wall is more than 5 billion m^2. The total area can be utilized more than 49 billion m^2. If the fixed PV area of architecture has a share of 20%, the capacity total is 100GW. Some actual applications of architecture PGEP have been implemented, such as solar energy demonstration city in Baoding, the international flower garden in Shenzhen and the Olympic Games gymnasium in Beijing.

Fortunately, the GPG have been regarded by the central government and some corporations. The biggest GPG in China lie in Dunhuang, Gansu Province. The total capacity is 10MW and the total investment is more than 73 million dollars and the area of PV is about 1 million m^2 and the annual accumulative total of electric power is about 16 million kWh. The item has a short transmission distance, and the distance is about 13km from Dunhuang City, which can provide clear energy for common people of Dunhuang. Certainly, some actual examples have been used to improve the energy structure, such as the total capacity of Chongming Island item in Shanghai is 1MW and the total capacity of Eerduosi item in Inner Mongolia is 255kW. So the GPG in China has a beautiful future with the increasing regard by the central government and common people.

At present, the PV generation of the whole world has a little share in the total of electric power system. According to the forecast of Europe Joint Research Center (JRC), with the increasing price of traditional energy, the energy structure of whole world will change in future. It is estimated that the renewable resources in 2030 has a share of 30% in the whole energy supply, and the PV generation has a share of 10% in 2030. The renewable resources in 2040 have a share of 50% in the whole energy supply, and the PV generation has a share of 20% in 2040. At the end of 21st century, the renewable resources have an incredible share of 80%, and the PV generation has a share of 60%. As mentioned above, the energy impact of China is more austere than the impact of world in the future. In order to settle the austere impact of economic and society sustainable development it is important to increase the share of PV generation in the whole energy supply. At present, PV generation in China has a share of 0%. Fortunately, the government has realized the importance of PV generation, and some intending objects have been established in the strategic programming of Chinese renewable resources exploitation from 2006 to 2020. According to the data of Chinese Development and Innovation Committee, the object of renewable energy development in 2020 contains: the large water electric power is 0. 3 billion kW, wind energy is 30GW, solar energy PV generating system is 1. 8GW, the biology energy is 30GW, solar water heater is 0. 3 billion m^2 and the biology fuel is 15 billion liters. In 2050, the renewable energy has a share of 25% in the whole energy supply, and the PV generation has a share of 5%. The capacity of PV generation is 100GW in 2050. The prospect of PV is enormous in future. The developmental speed of renewable resources is rapid in future China.

1.5.2.2 Situation about China solar cell industry

Throughout China, solar cells are held monthly seminars and exhibition. Moreover, the site also released the industry's key players a wealth of information and corporate executives.

On solar power, in 2009 the formation of the domestic market has to some extent. Ningxia is only one place to build a 40MW large power station. The results show that the factors affecting the cost structure in addition there are other modules. Currently, construction, management and operation cost is also high. Only solar cell technology (conversion efficiency, etc.) made great progress, in order to reduce costs to bring space. Detailed analysis must be thorough when needed to solve the problem. These are the systems and construction point of view. To solve these problems, but also take some time.

In 2009, the Chinese government's energy policy for the solar cell industry's recovery and development. Import capacity of wind power remains the top (as at September 2009 to 1600MW), solar cell production as of September 2009 to 2.5GW. In the utilization of new energy, solar power is a direct use of solar energy, can be said that the greatest agricultural revolution after the revolution.

In the Chinese domestic market, in 2009, set volume 160MW (yield of about 3.75%), China's domestic total to set more than 300MW, a year ahead to achieve the objectives of the Eleventh Five Year Plan. For the early implementation of the feed – in tariffs for the publicity. Strive to which the rapid formation of the domestic market. In solar power generation, power – storage technology could be a breakthrough if the energy can be to challenge the traditional fossil energy sources.

Solar cell production in China will expand under the government support. In 2009, global solar cell production reached 10.5GW, an increase of 32%. Set of solar power capacity 6.5GW, an increase of 18%. China's solar cell production reached 4GW, an increase of 50%, more crystalline silicon material production reached 20,000 tons, an increase of 225%.

As technology advances and the expansion of industrial scale, solar power costs can be greatly reduced. Some companies when the cost of power generation in 2006 was 4 yuan / (kW • h), while the end of 2009 came down to 1 yuan / (kW • h). Solar cells but also the cost of business will be down to 0.7 yuan / (kW • h). Originally it was envisaged to take 30 years, but the actual use of the 3 years to achieve.

Exercises and Discussion

1. What do you think of the solar PV situation in China?
2. Please describe the mechanism of the photovoltaic effect.
3. What do you think of the solar PV application relevant to the present market and explain why it will be the major new energy in future around 2050.
4. What are the utilization patterns for solar energy?

Chapter 2

Basic Knowledge of Solar Cells and Photovoltaic

2.1 P – n junctions of the solar cells

At the heart of solar cells is p – n junction. P – n junction results from the "doping" that produces conduction – band or valence – band selective contacts with one becoming the n – side (lots of negative charges), the other the p – side (lots of positive charges), that is, a p – n junction is formed by joining n – type and p – type semiconductor materials, as shown in Figure 2 – 1.

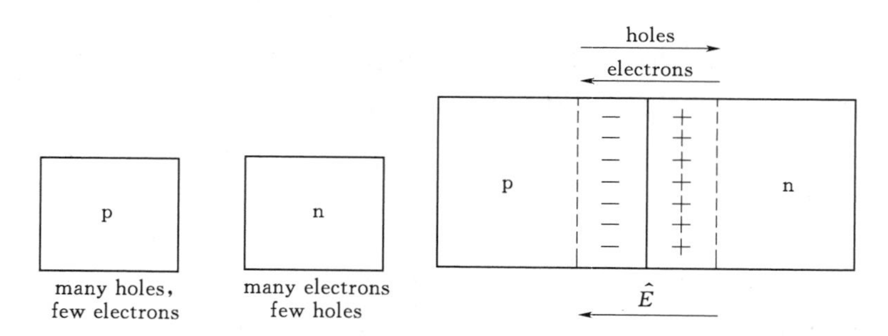

Figure 2 – 1 The formation of a p – n junction.
(Source: Stuart R. Wenham, Martin A. Green, Muriel E. Watt, Richard Corkish, Alistair Sproul, *Applied Photovoltaics*.)

When the n – and p – type materials are joined, the excess holes in the p – type material flow by diffusion to the n – type material, while electrons flow by diffusion from the n – type material to the p – type material as a result of the carrier concentration gradients across the junction. The electrons and holes leave behind exposed charges on dopant atom sites, fixed in the crystal lattice. An electric field ($\hat{E}$) therefore is built up in the so – called depletion region around the junction to stop the flow. Depending on the materials used, a "built – in" potential (V_{bi}) owing to $\hat{E}$ will be formed. If a voltage is applied to the junction, as shown in Figure 2 – 2, $\hat{E}$ will be reduced.

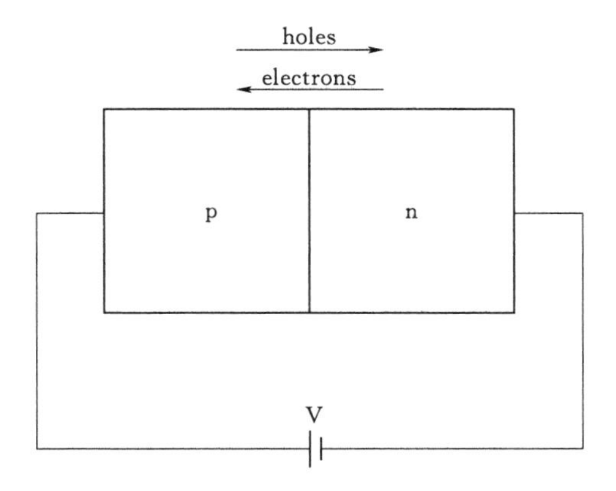

Figure 2 - 2 Space charge region of the p - n junction.
(Source: Stuart R. Wenham, Martin A. Green, Muriel E. Waft, Richard Corkish, Alistair Sproul, Applied Photovoltaics.)

Specialized Vocabulary:

- p - n junction p - n 结
- doping (在半导体材料中) 掺杂
- conduction - band 导带、传导带、导电带
- valence - band 价带
- negative charge 负电荷
- positive charge 正电荷
- carrier 载流子
- depletion region 耗尽区，势垒区，阻挡层
- built - in potential 内建电势，内建电场，内建电位
- electric field 电场

2.2 Solar cell types

The first generation is the silicon based solar cells. It has the most mature technology and has been developed for the longest time. The first generation of solar cells includes monocrystalline silicon, polycrystalline silicon, amorphous silicon solar cells. The second generation is the inorganic compounds thin - film solar cells such as Cadmium Telluride (CdTe), Copper Indium Selenide (CIS), Copper Indium Gallium Selenide (CIGS), Gallium Arsenide (GaAs), etc. The third generation is the new concept solar cells, such as organic solar cell/organic photovoltaics (OPV), dye - sensitized solar cells (DSSC), perovskite solar cells (PSC), quantum dots solar cells (QDSC), etc. The big difference between the third generation solar cells and the other solar cells is, some introduce organics and nanotechnology into the production process, and do not necessarily have a minority carriers but develop the PV effect through kinetics of charge transfer.

There are different ways of classifying existing solar cells.

(1) According to the base material, we can divide the solar cells into four types. ①Crystalline Silicon Solar Cells, including Monocrystalline silicon and Polycrystalline silicon. ② Amorphous Silicon Solar Cells. ③ Inorganic Compound Solar Cells, including GaAs, CdS, CdTe, CIGS, etc. ④ Organic & Nanotechnology Semiconductor Solar Cells, including DSSC, OPV, PSC, QDSC.

(2) According to the structure of the solar cells, there are five types of solar cells. ①Homojunction solar cells, including silicon solar cells, GaAs cells. ② Heterojunction solar cells, such as SnO_2/Si, Si/GaAs. ③Schottky junction solar cells, such as metal - semiconductor. ④Composite/Multi - junction solar cells, including multiple p - n juncions. ⑤Liquid junction solar cells, such as photoelectrochemical solar cell.

(3) According to the application of the solar cells, there are solar cells applied in space, Terrestrial solar cells, and Light sensors.

(4) According to the work mode, there are flat solar cells and concentrator solar cells.

Specialized Vocabulary:

- monocrystalline silicon 单晶硅
- polycrystalline silicon 多晶硅
- amorphous silicon 非晶硅
- Inorganic compounds thin - film solar cells 无机化合物薄膜太阳电池
- organic solar cells/organic photovoltaics (OPV) 有机太阳电池
- dye - sensitized solar cells (DSSC) 染料敏化太阳电池
- perovskite solar cells (PSSC) 钙钛矿太阳电池
- quantum dots solar cells (QDSC) 量子点太阳电池
- Light sensors 光感应器
- flat solar cells 平板太阳电池
- concentrator solar cells 聚光太阳电池

2.3 The structure and working mechanism of solar cells

The appearanceand the basic structure of the silicon solar cell are shown in Figure 2 - 3. The basic material is the p - type monocrystalline silicon with the upper surface which is n+ type area, together is p - n junction. There is metal palisade electrode (structure to allow light to enter the cell) on the top surface, and the metal bottom electrode is on the back of the silicon. Top and bottom electrodes contact with n area and p area respectively to form ohmic contacts. There is also the antireflection film covered evenly on the whole top surface.

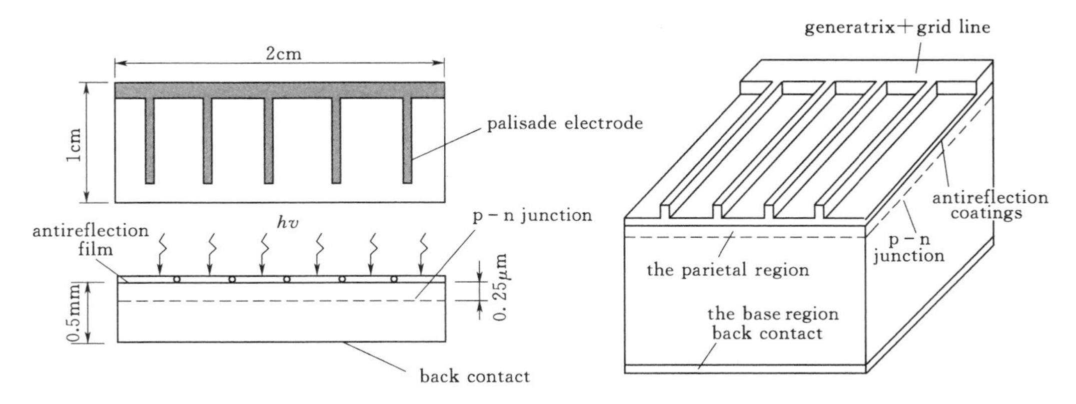

Figure 2 - 3 Basic structures of the silicon solar cell and monocrystalline silicon solar cell.

The specific components of monocrystalline silicon solar cells are described below:

(1) The back ohmic contact area. The metal and the base region (single crystal substrate) form ohmic contact at the back surface.

(2) The base region. The base region consists of single crystal substrate materials. The function of the substrate is to support the solar cells. The substrate for single crystalline solar cell is silicon wafer, for thin film solar cells it can be glass, metal foil or flexible materials, etc. (Figure 2 - 4)

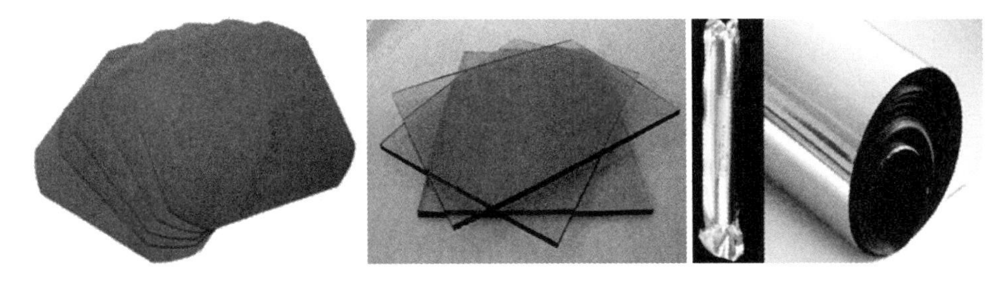

Figure 2 - 4 Different substrates: silicon wafer, glass, metal foil.

(3) The junction area, namely, p - n junction barrier area. Both sides of the junction can be made of the same material or different materials. If the p - n junction is composed of the same semiconductor materials (only different conductive types), it is known as the homojunction. If the p - n junction is made of different materials, it is known as the heterojunction. Generally, single crystal solar cells are using homojunction, the thin film solar cells which were developed more recently are using heterojunction. In addition, in the single crystal and thin - film solar cells, the Schottky barrier area also has been used.

The p - n junction is used to generate the photovoltaic effect. The function of the p - n junction is absorption and conduction. In solar cells, the absorption of photons occurs at the p - type layer. It is important to have better absorption ratio and generate more conduction electrons in the p - type layer. Another important goal is reducing the recombination of electrons and holes to extend the charge lifetime of the solar cells.

(4) The top region. Another layer of semiconductor is forms the top region which is in contrast to the material of the conductive area, the top region is much thinner than the base region. In the case of the front illumination, most of the light (especially the long wavelength photon) crosses the top region and the junction area to reach the base region, and it is absorbed within the thikness of the base region which is closed to the p – n junction, and then produces electron – hole pairs. Most of the minority carriers diffuse and drift to the parietal area to form the photocurrent, and the recombination of a few minority carriers happened in the diffusion process.

(5) The metal grid contact (including grid line and the generatror) . They covered about 5%～10% of the photosensitive area. The metal electrode in the solar cells is used as electric contact (Figure 2 – 5) . There are back electrode and front electrodes. Generally, Aluminum (Al) or Molybdenum (Mo) are used as the back electrode. The front electrodeis, normally in gratings pattern in order to reduce shading losses. Physical vapor deposition (PVD) and chemical vapor deposition (CVD) have been used to deposit metal electrode to achieve good performance, but it has high cost.

Figure 2 – 5 Metal electrode in gratings pattern.

(6) Antireflection coating (ARC) . The purpose of the antireflection coating is to reduce the reflection loss of the incident light, increase the proportion transmitted part and increase the photocurrent. The ARC materials commonly used are TiO_2, SiO_2, Ta_2O and ITO (indium tin oxide), etc. For silicon based solar cells, silicon is shiny gray and it can reflect up to 30% of the incident light, which will reduce energy transfer efficiency. After coating with the anti – reflection layer which has a texture surface structure, the improvement of the incident light capture will go up to 10%～15% (Figure 2 – 6) .

(7) Encapsulation. The commonly used materials for encapsulation are glass and cellophane tape.

(8) If it is the thin film solar cell, the above structure will all be deposited on a transparent glass or other cheap substrate.

These are some performance requirements of the materials used for solar cells. The bandgap of the materials is between 1.1eV and 1.7eV, and it is preferably a direct bandgap material, non – toxic material, able to be produced in large areas, with high photovoltaic conversion efficiency and long term stability.

The power generation principle of the solar cell is based on the photoelectric effect

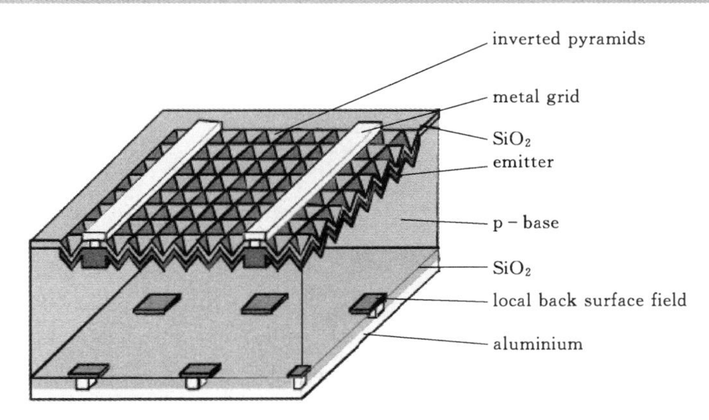

Figure 2 - 6 Surface textures as anti - reflection layer for solar cells.

caused by the light irradiating on the semiconductor. In irradiation, the energy of the photons is larger than the forbidden band width which will cause production of electrons and holes. The p - n junction in the semiconductor will allow the electrons to diffuse to the n - type semiconductor and the holes diffuse to the p - type semiconductor, namely, the negative charges and the positive charges will gather at the both sides, respectively. Therefore, when these two electrodes are connected, the charges will flow to generate electricity. This is completely different from the traditional power generation, in producing DC power, not have the rotating parts from the motor, nor exhaust gas, so it is a clean, noiseless electric generator.

Specialized Vocabulary:

- ohmic contact 欧姆接触
- antireflection film 减反射膜
- antireflection coating (ARC) 抗反射层
- base region 基区、基极区
- homojunction 同质结
- heterojunction 异质结
- The junction area 结区
- The top region 顶区
- electron - hole pairs 电子—空穴对
- the parietal area 顶叶区
- The metal grid contact 金属栅接触
- grid line 栅线
- generatror 发电机、发生器、产生器
- forbidden band 禁带
- physical vapor deposition (PVD) 物理气相沉积
- chemical vapor deposition (CVD) 化学气相沉积

2.4 The fundamental characteristics of the solar cells

The fundamental characteristics of the solar cells are the polarity of the solar cell, the performance parameters of the solar cell, and volt - ampere characteristics are as follows.

(1) The polarityof the solar cell. The silicon solar cell is generally made in p+/n type structure or n+/p type structure. p+ and n+ mean the conduction type of the conductive materials from the front illumination layer of the solar cell. N and p mean the conduction type of the substrate conductive materials from the back side of the solar cell. The electrical properties of the solar cell is related to the characteristics of the semiconductor materials used for the solar cells.

(2) The performance parameters of the solar cell. These include open - circuit voltage, short - circuit current, maximum power output, fill factor and transfer efficiency, etc. These parameters are the data to evaluate if the solar cell performance is good or bad.

(3) The volt - ampere characteristic of the solar cell. Generally, the solar cell module, a multimeter and load resistance are used to create the circuit, and the load resistance R is changed, measuring the current and the voltage drop across the load, then graphing the volt - ampere characteristic curves of the PV module.

Specialized Vocabulary:

- polarity 极性
- volt - ampere characteristic 伏安特性
- open - circuit voltage 开路电压
- short - circuit current 短路电流
- maximum power output 最大输出功率
- fill factor 填充因子
- (electron) transfer efficiency 转换效率
- solar cell module 太阳电池组件
- load resistance 负载电阻
- multimeter 万用表

2.5 Technology trends

The trend of the solar cell technology in future will include flexible, thin - film solar cells with the characteristic of low - cost, high efficiency and long life, multi - junction tandem cells such as GaInP/GaAs/New/Ge for new materials are the challenge, and concentrator solar cells. Future development goals are to improve the conversion efficiency, reduce manufacturing costs and expand the application field.

The second technology trend is grid - connected PV power plant with extra large capacity (100MW and GW level) plus grid - connected PV power plant integration technology, developing large - scale grid - connected inverter and efficient cost effective concentrator photovoltaic power generation system.

The third technology trend is BIPV. In the future, people will expect to use new photovoltaic building materials and building components with access for integrated PV, with case of operation and a standard photovoltaic price.

2.6 Reading material (Please read the article and find out the specialized vocabulary)

The Photovoltaic (PV) Effect

(Basic photovoltaic principles and methods - Published by Technical Information Office, 1982, Solar Information Module 6213, SERI/SP - 290 - 1448)

HIGHLIGHTS

The photovoltaic (PV) effect is the basis of the conversion of light to electricity in photovoltaic, or solar, cells. Described simply, the PV effect is as follows: Light, which is pure energy, enters a PV cell and imparts enough energy to some electrons (negatively charged atomic particles) to free them. A built - in - potential barrier in the cell acts on these electrons to produce a voltage (the so - called photovoltage), which can be used to drive a current through a circuit.

This description does not broach the complexity of the physical processes involved. Although it is impossible here to cover fully all the phenomena that contribute to a PV - generated current, it is possible to go deeply enough into these phenomena to understand howan effective cell works and how its performance can be optimized. We can do this by answering some fundamental questions about processes central to the working of a PV cell:

1. What does it mean to say that an electron is freed?

a. Where is it freed from?

b. Where does it go?

2. What is the potential barrier that acts on the free electrons?

a. How is it formed?

b. What does it do?

3. Once acted on by the potential barrier, how do the free charges produce a current?

We shall take the silicon cell as a model. Silicon is a widely used, typical cell material; understanding the silicon cell is a good groundwork for understanding any PV cell. We shall start by reviewing some of silicon's basic atomic characteristics.

AN ATOMIC DESCRIPTION OF SILICON

All matter is made from atoms. They, in turn, are composed of three kinds of particles:

protons, neutrons, and electrons. Protons (positively charged) and electrons (negatively charged) attract each other; neutrons are not electrically attracted to either and are said to be neutral. The positively charged protons and the neutral neutrons reside in a nucleus, the close - packed center of the atom. The electrons - much lighter than the protons (or neutrons) - orbit the nucleus. Although an atom contains charged particles, overall it is electrically neutral because it has the same number of protons and electrons.

Different atoms have different numbers of protons. For every proton in an atom's nucleus, there is an electron orbiting the nucleus. The orbital locations (and the motion of the 'electrons about their own axis) are determined by the energy of the electrons. The electrons, in particular those furthest from the nucleus, interact with electrons from other atoms and determine the way in which like or dissimilar atoms combine into larger structures such as solids.

The silicon atom has fourteen electrons arranged in such a way that the outer four can be given to, accepted from, or shared with another atom. These four outer electrons are called valence electrons.

Large numbers of silicon atoms, through their valence electrons, can bond together to form a solid. As a solid, each silicon atom usually shares each of its four valence electrons with another silicon atom. Each basic silicon unit, forming a tetrahedral arrangement, thereby contains five atoms (the one silicon atom plus the four others it shares electrons with).

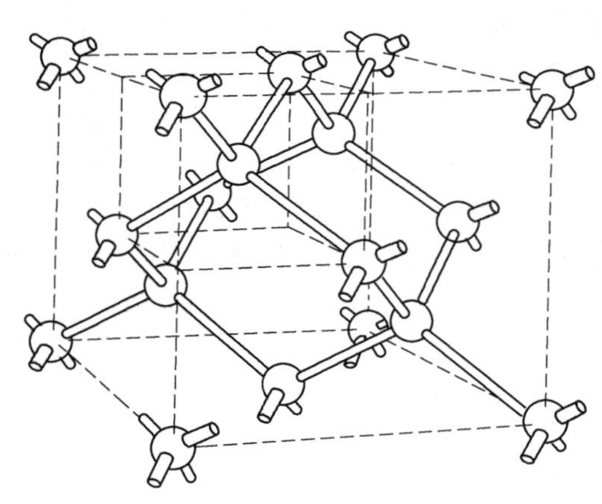

Figure 2 - 7 Representation of the silicon crystal lattice arrangement.

Each atom in the silicon solid is held in place at a fixed distance and anglewith each of the atoms with which it shares a bond. This regular, fixed formation of a solid's atoms is called a crystal lattice.

Solids can form from several differently shaped crystal lattices. (All solids are not crystalline, however; some can have multiple crystalline forms and/or none at all.) For silicon (Figure 2 - 7) the atoms are located so as to form the vertices of a cube with single atoms centered at each of the faces of the cubic pattern. The cubic arrangement repeats throughout the crystal.

THE EFFECT OF LIGHT ON SILICON

When light strikes a silicon crystal, it may be reflected, be absorbed, or may go right through. Let's concentrate on the light that is absorbed. Usually when light of relatively low energy is absorbed by a solid, it creates heat without altering the electrical properties of the material. That is, low - energy light striking a silicon crystal causes atoms of silicon to vibrate and twist in their bound positions, but do not break loose. Similarly, electrons in bonds also gain more energy and are said to attain a higher energy level. Since these energy levels are not stable, the electrons soon return to their original lower energy levels, giving off as heat the energy they had gained.

Light of greater energy can alter the electrical properties of the crystal. If such light strikes a bound electron, the electron is torn from its place in the crystal. This leaves behind a silicon bond missing an electron and frees an electron to move about in the crystal. A bond missing an electron, rather picturesquely, is called a hole. An electron free to move throughout the crystal is said to be in the crystal's conduction band (Figure 2 - 8), because free electrons are the means by which electricity flows. Both the conduction - band electrons and the holes play important parts in the electrical behavior of PV cells. Electrons and holes freed from their positions in the crystal in this manner are said to be light - generated electron - hole pairs.

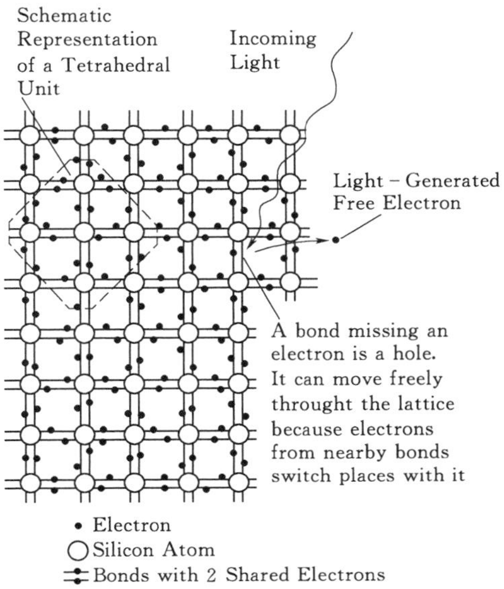

Figure 2 - 8 Light of sufficient energy can generate electron - hole pairs in silicon, both of which move for a time freely throughout the crystal.

A hole in a silicon crystal can, like a free electron, move about the crystal. The means by which the hole moves is as follows: An electron from a bond near a hole can easily jump into the hole, leaving behind an incomplete bond, i. e. , a new hole. This happens fast and frequently - electrons from nearby bonds trade positions with holes, sending holes randomly and erratically throughout the solid. The higher the temperature of the material, the more agitated the electrons and holes and the more they move.

The generation of electrons and holes by light is the central process in the overall PV effect, but it does not itself produce a current. Were there no other mechanism in volved in a solar cell, the light - generated electrons and holes would meander about the crystal randomly for a time and then lose their energy thermally as they returned to valence positions. To exploit the electrons and holes to produce an electric force and a current, another mechanism is needed - a built - in "potential" barrier.

THE POTENTIAL BARRIER

The Function of the Barrier

A PV cell contains a barrier that is set up by opposite electric charges facing one another on either side of a dividing line. This potential barrier selectively separates light - generated electrons and holes, sending more electrons to one side of the cell, and more holes to the other. Thus separated, the electrons and holes are less likely to rejoin each other and lose

their electrical energy. This charge separation sets up a voltage difference between either end of the cell (Figure 2 - 9), which can be used to drive an electric current in an external circuit.

Forming the Barrier

There are several ways to form a potential barrier in a solar cell. One is to slightly alter the crystal so that its structure on either side of the dividing line is different.

The Negative - Carrier (Donor) Dopant. As previously indicated, silicon has four valence electrons, all of which are normally part of bonds in a silicon crystal. Suppose by some means we introduce an impurity into an otherwise pure silicon crystal by' substituting for a silicon atom an atom such as phosphorus, having five valence electrons. The impurity atom would occupy the same position in the crystal as a normal silicon atom, supplying an electron for each of silicon's four bonds. But because the phosphorus atom has one extra valence electron, there would be one electron with no bond to share (Figure 2 - 10) . Compared with a bound electron, the impurity atom's extra electron is relatively free. In fact, at room temperature there is enough thermal energy in the crystal to shake this electron loose, despite the fact that it would leave behind a positively charged impurity atom. This free electron from the impurity has no hole (empty bond) into which it may readily drop, and it behaves as if it were a permanent member of the crystal's conduction band, always ready to be part of an electric current.

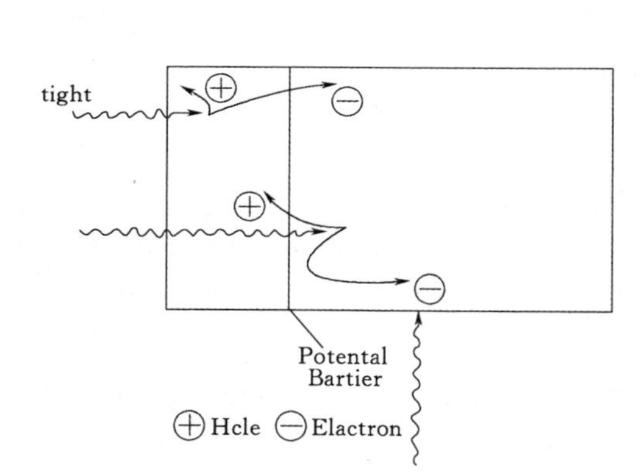

Figure 2 - 9 A potential barrier in a solar cell sepa - rates light - generated charge carriers, creating a voltage.

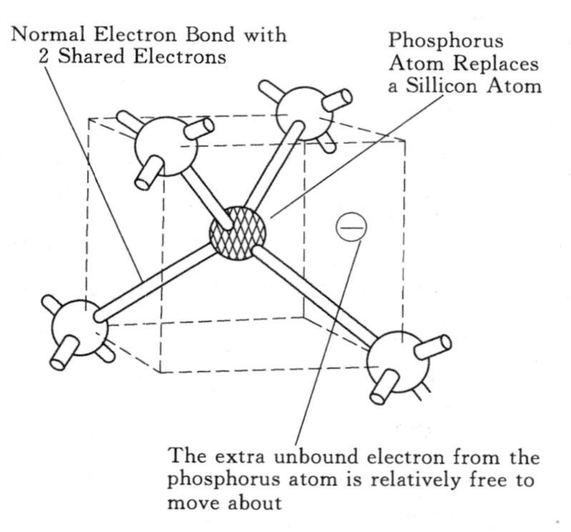

Figure 2 - 10 When an impurity atom such as phos - phorus with five valence electrons is substituted into a silicon crystal, it has an extra, unbonded electron.

A silicon crystal with numerous substituted phosphorus atoms would have many free, conduction band electrons and a similar number of positive impurity ions locked into the crystal's structure. Overall, the whole crystal would remain neutral, since there are just as many positive ions as free electrons; but the crystal's electrical properties would have been drastically altered.

Impurities introduced in this way are called dopants, and dopants that have one extra valence electron (such as phosphorus introduced into a. silicon crystal) are called donors because they donate an electron to the crystal. Such a donor - doped crystal is known as n - type because it has free negative charges.

Altering silicon by introducing a donor dopant is part of the process used to produce the internal potential barrier. But n - type silicon cannot of itself form the barrier; other, slightly altered silicon is also necessary, this kind with electrical properties opposite those of the n - type silicon.

The Positive - Carrier (Acceptor) Dopant. An appropriately altered material can be formed by substituting into the silicon crystal, impurity atoms with one fewer valence electron than silicon. An impurity atom with three valence electrons (such as boron) would sit in the position of the original silicon atom, but one of its bonds with the silicon would be missing an electron, i. e., there would be a hole (Figure 2 - 11). As we saw before, holes can move about almost as freely as conduction - band electrons. In this way, a silicon crystal doped with many such boron atoms has many holes that act as if they were free positive charges moving throughout the crystal lattice.

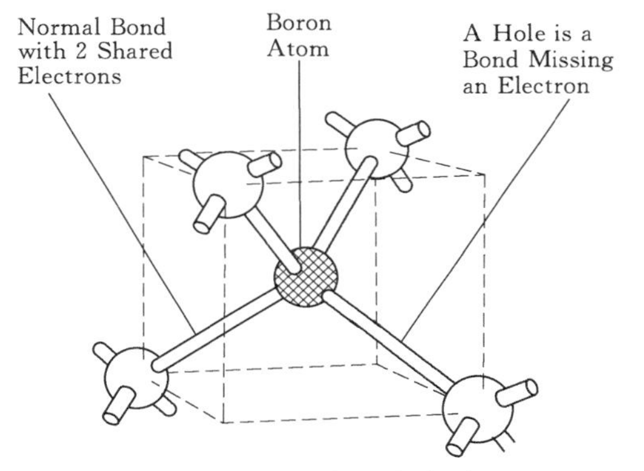

Figure 2 - 11 A three - valence - eletron impurity (such as boron) in a silicon crystal is normally bonded, except one of the bonds is missing an electron, i. e., is a hole.

A three - valence - electron impurity in a silicon crystal is called an acceptor because its holes accept electrons (normally bonded valence electrons or conduction band electrons) from the rest of the silicon crystal. An acceptor - doped silicon material is called p - type because of the presence of free positive charges (the moving holes).

In a p - type material, positive charges are the so - called majority (charge) carriers because they far outnumber any free electrons that in p - type materials are referred to as minority carriers. In an n - type material, where the doping is reversed, electrons (negative charges) are the majority carriers and holes the minority carriers.

The Junction. A line dividing n - type from p - type silicon establishes the position of a potential barrier essential to the operation of a solar cell. To see how this barrier comes into being, let's take a look at the junction between the two materials (the area in the immediate proximity of the two surfaces). In the p - type material, there are excess holes; in the n - type material, excess electrons [Figure 2 - 12 (a)]. When the n and p materials are in contact, free electrons in the n - type

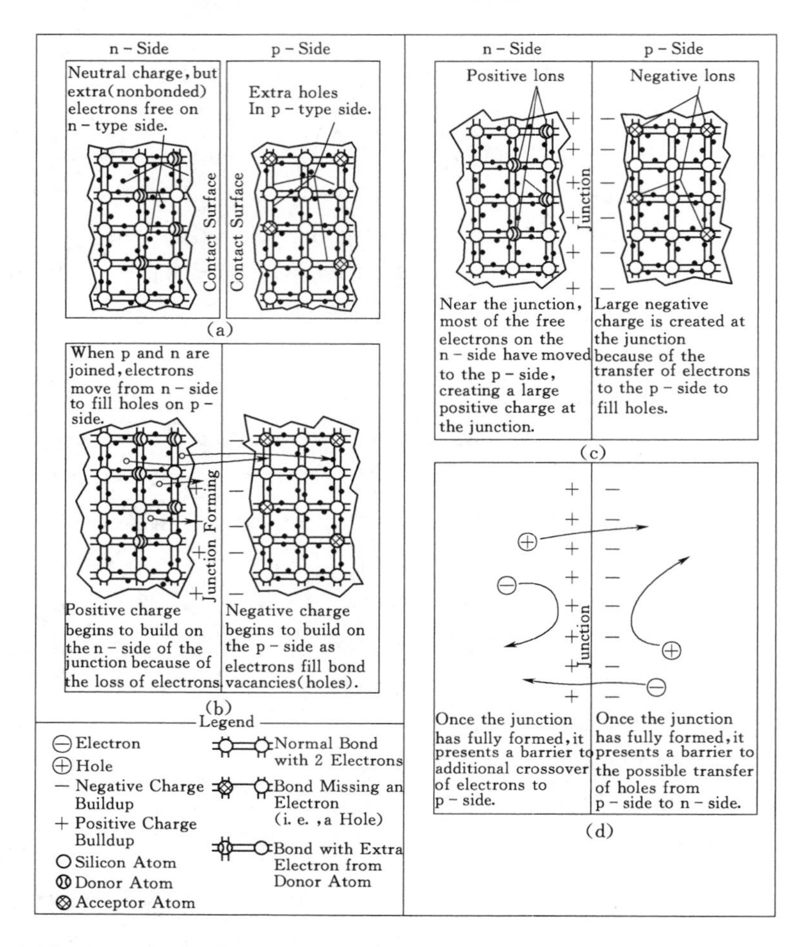

Figure 2 - 12 During junction formation, electrons move from the n - type silicon into the p - type, while holes move in the opposite dirction. Movement of electrons into the p - type silicon and holes into the n - type silicon bullds up a fixed potential barrier at the junction opposing the further movement of free carriers and creating a state of equilibrium.

material adjacent to the many holes in the p - type material at the junction will jump into the p - type material, filling the holes. Also, valence band electrons on the n - type side can jump into holes on the adjacent p - type side, which is equivalent to a hole moving over into the n - type material (for simplicity, this is not shown in Figure 2 - 12) . This charge transference process happens rapidly along the dividing line (junction), sending huge numbers of electrons to 'the p - type side and holes to the n - type side [Figure 2 - 12 (b)] . This causes an immediate imbalance of charge: more negative charges (extra electrons) along the p - type side of the interface, more positive charges (ions) along the n - type side [Figure 2 - 12 (c)] .

When electrons move over into the p - type material during junction formation, they find holes in the silicon bonds and drop into them. In like manner, holes that transfer to the n - type side are quickly filled by the n - type side's numerous extra electrons. Consequently, carriers that form the junction lose their freedom of movement. Thus, although a charge imbalance exists at the junction, there are very few free electrons on the p-type silicon side to be pulled back to the n-type side, and very few free holes on the n-type side to be transferred back to the p-type material. So, the charge imbalance stays fixed in place.

The Barrier. The process of charges moving across the junction to create a charge imbalance in the above described manner does not continue indefinitely. Charged carriers that have already crossed the junction set up an electric force (field) that acts as a barrier opposing the further flow of free carriers. As more carriers cross the junction, the barrier enlarges, making it increasingly difficult for other carriers to cross. Eventually, an equilibrium is established where (statistically speaking) no more electrons or holes switch sides. This creates a fixed potential barrier at the junction (the barrier to which we have been referring since the beginning), with the n-type side adjacent to the junction being positively charged and the p-type side adjacent to the junction being negatively charged. The "height" (that is, the strength of the electric force) of the barrier, it should be noted, depends upon the amount of dopant in the silicon - the more the dopant, the more charge imbalance induced and the greater the barrier's ability to separate charges.

Let us note some qualities of the barrier [Figure 2-12 (d)]. It opposes the crossing of majority charge carriers. That is, electrons in the n-type material would have to climb the barrier against the built-in field to enter the p-type material. Similarly, holes in the p-type region are held back from entering the n-type region. Note also that minority carriers are not hindered by the barrier. In fact, free electrons on the p-type side - of which there are very few, being the minority carrier there - are driven by the junction field to the opposite, n-type side. The same is true of holes driven from the n-type side. But normally (under no illumination) there are so few minority carriers on their respective sides that their movement is nil; and what there is, is balanced by the few majority carriers that randomly assume enough energy to cross the barrier. This selective barrier at the junction is the means of separating charges during electron-hole generation under illumination. It is the key to the production of a PV electric current.

The Potential Barrier in Action

For illustrative purposes, suppose light striking the PV cell has enough energy to free an electron from a bond in the silicon crystal. This creates an electron-hole pair - a free electron and a free hole. Suppose further that the electron-hole pair is generated on the p-type silicon side of the junction. An electron from such an electron-hole pair has only a relatively short time during which it is free because it is very likely to combine with one of the numerous holes on the p-type side. But solar cells are designed so that in all probability the electron will meander around the crystal and encounter the junction before it has the chance to combine with a hole (Figure 2-13). (Were it to combine with a hole, it would lose its energy as heat and be useless as far as PV electric current is concerned.)

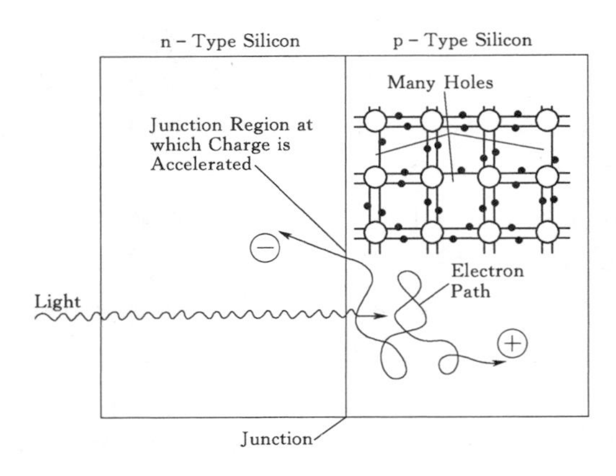

Figure 2 - 13 A photogenerated electron on the p - type side usually has enough time to bounce randomly around the crystal and encounter the junction before it can combine with a hoie.

Once the free electron is within the field of the junction (which is limited to the junction's immediate vicinity), the electron is accelerated across the barrier (by the barrier's charge imbalance) into the n - type silicon. Since there are very few holes on the n - type side of the junction, the electron is no longer in great danger of recombining. Moreover, there is very little chance of its returning to the p - type side because it would have to buck the repulsion of the junction's field (climb the barrier), expending energy it usually doesn't have.

The hole partner of this electron - hole pair, however, remains on the p - type side of the junction because it is repelled by the barrier at the junction. It is not in danger of recombining because there are already a predominance of holes on the p - type side.

A similar situation occurs when the electron - hole pairs are generated by light on the n - type side of the junction. This time the freed electrons remain on the n - type side, being repelled by the barrier. Meanwhile, most of the holes encounter the junction before chancing to recombine. They cross the junction into the p - type side when normally bound electrons from the p - type side jump the junction and fill the holes. Once on the p - type side, the holes move around unhindered, and there are very few free electrons available to fill them.

Because illumination and charge separation causes the presence of uncombined excess negative charges on the n - type side and excess holes on the p - type side, a charge imbalance exists in the cell.

THE ELECTRIC CURRENT

If we connect the n - type side to the p - type side of the cell by means of an external electric circuit, current flows through the circuit (which responds just as if powered by a battery) because this reduces the light - induced charge imbalance in the cell. Negative charges flow out of the electrode on the n - type side, through a load (such as a light bulb), and perform useful work on that load (such as heating the light bulb's filament to incandescence). The electrons then flow into the p - type side, where they recombine with holes near the electrode (Figure 2 - 14). The light energy originally absorbed by the electrons is used up while the electrons power the external circuit. Thus, an equilibrium is maintained: The incident light continually creates more electron - hole pairs and, thereby, more charge imbalance; the charge imbalance is relieved by the current, which gives up energy in performing work.

The amount of light incident on the cell creates a near - proportional amount of current.

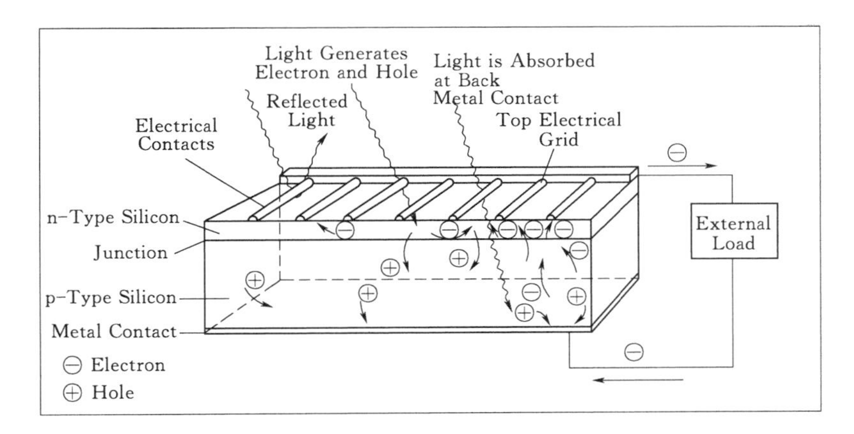

Figure 2 - 14 Light incident on the cell creates electron - hole pairs, which are separated by the potential barrier, creating a voltage that drives a current through an external circuit.

The amount of energy it takes to raise an electron to the conduction band is the amount of energy the light originally imparts to the electron and is, thus, the maximum that can be retrieved from the electron in the external circuit.

We have observed all the conditions necessary for current to flow: incident light to free the charge carriers, a barrier to accelerate the carriers across the junction and keep them at opposite ends of the cell, and a charge imbalance to drive a current (charged carriers) through a circuit.

Exercises and Discussion

1. Whatis p - n junction?
2. What kind of characterizations that we can use for solar cells?
3. Please describe briefly about different types of solar cells.
4. What are the structure and mechanism for the silicon solar cells?

Chapter 3

Silicon Solar Cells

3.1 Semiconductor types

The atomic structure of silicon makes it one of the ideal elements for photovoltaic of solar cells. The silicon atom has 14 electrons and its structure is such that its outermost electron shell contains only four electrons. In order to be stable, this shell needs to have eight electrons. In its normal state or pure form, each silicon atom attaches itself to four other silicon atoms to form a stable silicon crystal. In the silicon crystal's pure state, there are very few free electrons available for carrying the electric current. In order to alter their electrical conductivity, other elements are introduced to the silicon as impurities (dopants) in a process known as doping. Moreover, the silicon is non-toxic, abundant in the world, relatively cheap and a mature commodity.

Crystalline silicon has a fundamental indirectband gap of $E_g = 1.17eV$ and a direct band gap above 3eV [4] at ambient temperature. These characteristics determine the variation of optical properties of Si with wavelength, including the low absorption coefficient for carrier generation of near band gap photons [5]. At short ultraviolet (UV) wavelengths in the solar spectrum, the generation of two electron-hole pairs by one photon seems possible, though quantitatively this is a small effect [6]; at the other extreme of the spectrum parasitic free-carrier absorption competes with band-to-band generation [7]. The intrinsic concentration is another important parameter related to the band structure; it links carrier disequilibrium with voltage [8].

Crystalline silicon solar cells and modules have dominated PV technology from the beginning. They constitute more than 85% of the PV market today, and although their decline in favor of other technologies has been announced a number of times, they presumably will retain their leading role for some time, at least for the next decade. One of the reasons for crystalline silicon to be dominant in PV is the fact that microelectronics has developed silicon technology greatly. On one hand, not only has the PV community benefited from the accumulated knowledge but also silicon feedstock and second-hand equipment have been acquired at reasonable prices. On the other hand, microelectronics has taken advantage of some innovations and developments made in PV.

Silicon and other semiconductor materials used for solar cells can use monocrystalline, multicrystalline/polycrystalline, microcrystalline or amorphous silicon materials. Although usages of these terms vary, there is the definition by planar grain size according to Basore (1994). Microcrystalline material has grains smaller than 1μm, polycrystalline smaller than 1mm and multicrystalline smaller than 10cm. The structure of the different material types is illustrated in Figure 3-1.

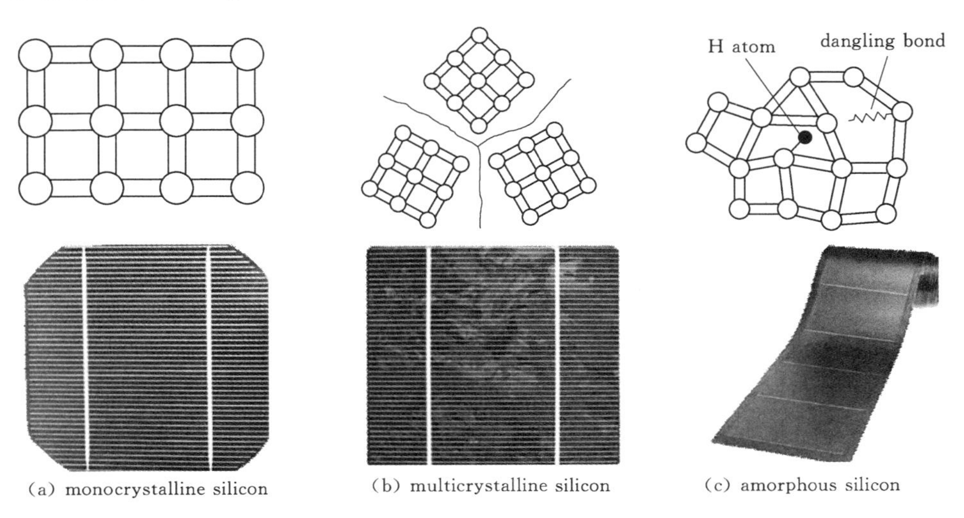

(a) monocrystalline silicon (b) multicrystalline silicon (c) amorphous silicon

Figure 3-1 Different structures and surface morphology of monocrystalline silicon.

As shown in Figure 3-1, the structure among different types of crystalline silicon is different. For the monocrystalline silicon (c-Si) material [Figure 3-1 (a)], the atoms are arranged in a regular pattern which have periodic and are ordered. Multicrystalline or polycrystalline silicon material [Figure 3-1 (b)] is composed of single crystals of silicon with different crystal orientation, irregular shape and grain boundaries, the regions of crystalline silicon are separated by "grain boundaries", where bonding is irregular, but the atoms in each silicon crystal are arranged in a regular pattern. For amorphous silicon material which has less regular arrangement of atoms [Figure 3-1 (c)], the atoms inside each "crystal" are disordered and unsystematic, this enable bending of the material free and it is "soft". This disorganized arrangement leads to "dangling bonds" in the silicon that can be passivated by hydrogen.

It has already been pointed out that the peculiarities of multicrystalline silicon cells may prevent, in some cases, the use of standard processing technologies. Some of the proposed alternatives are not yet sufficiently cost-effective as to be incorporated in an industrial production line, but others utilised. Two main differences with single crystalline silicon materials can be highlighted.

(1) Multicrystalline material quality is poorer because of crystalline defects (such as grain boundaries, dislocations) and metallic impurities (dissolved or precipitated), giving lower bulk

lifetimes and hence lower cell efficiencies. To address this problem, two main strategies are followed, implementation of gettering steps and defect passivation with hydrogen.

(2) Texturing is more difficult because of different exposed crystallographic planes, so that standard alkaline solutions are not appropriate. To improve light – trapping and absorption, other techniques have to be implemented.

From Figure 3 – 1, we can see that crystalline silicon solar cells derive their name from the way they are made. At solar panels level, the difference between monocrystalline and polycrystalline is that monocrystalline cells are cut into thin wafers from a singular continuous crystal that has been grown for this purpose. Polycrystalline cells are made by melting the silicon material and pouring it into a mould. The uniformity of a single crystal (mono – crystalline) cell gives it an even deep blue color throughout. It also makes it more efficient than the polycrystalline solar modules whose surface is "jumbled" with an appearance of various shades of blue. Apart from the crystal growth phase, there is little difference between the construction of mono – and polycrystalline solar cells. The cells are usually laminated using tempered glass on the front and plastic on the back. These are joined using a clear adhesive and then the module is framed with aluminum. Single crystal modules are smaller in size per watt than their polycrystalline counterparts.

Specialized Vocabulary:

- band gap 带隙、能隙、能带间隙
- indirect band gap 间接带隙
- absorption coefficient 吸收系数
- intrinsic concentration 本征浓度
- grain boundaries 晶界

3.2 Monocrystalline silicon solar cells

As mentioned above, monocrystalline silicon material has an ordered crystal structure (Figure 3 – 2), with each atom ideally lying in a pre – ordained position. It therefore allows ready application of the theories and techniques developed for crystalline material, and exhibits predictable and uniform behavior. It is, however, the most expensive type of silicon, because of the careful and slow manufacturing processes required. Therefore, the cheaper multicrystalline or polycrystalline silicon (poly – silicon), and to a lesser extent, amorphous silicon are increasingly being used for solar cells, despite their less ideal qualities.

Monocrystalline silicon cells are made using cells saw – cut from a single cylindrical crystal of silicon, they are effectively a slice from a crystal. The Czochralski (CZ) technique, Float Zone (FZ) method and epitaxial method are used to produce single crystalline

Figure 3-2 Crystal structure of the monocrystalline silicon material.

silicon. The principle advantage of monocrystalline cells are their high lab efficiency over 25%, although the manufacturing process required to produce monocrystalline silicon is complicated, resulting in slightly higher costs than other technologies. In appearance, it will have a smooth texture and be able to see the thickness of the slice. They are also rigid and brittle so must be mounted in a rigid frame and substrate top to protect them.

Monocrystalline cells were first developed in 1955. They conduct and convert the sun's energy to produce electricity. When sunlight hits the silicon semiconductor, enough energy is absorbed from the light to knock electrons loose, allowing them to flow freely. Monocrystalline silicon solar cells are designed in such a way that the free electrons can be directed, within the cell's electric field, in a path or a circuit as electricity which is then used to power various appliances . The power (measured in watts) of the cell is determined by the combination of the current and the voltage of the cell. The voltage depends on the cell's internal electric field.

The disadvantages are high temperature required to form the boule (ingot) of pure silicon, and energy intensive manufacturing process, use of a relatively large amount of Si, expensive, fragile and low band-gap (1.17eV $\approx$ 1060nm) .

Specialized Vocabulary:

- energy intensive manufacturing process 能源密集型制造过程
- texturing 织构化
- texture etch 织构刻蚀、表面刻蚀
- plasma etch 等离子体刻蚀
- sinter 烧结
- diffusion 扩散
- edge insulation processing 边缘绝缘处理
- Czochralski (CZ) technique 直拉法
- Float Zone (FZ) method 区熔法
- epitaxial method 外延法

3.3 Multicrystalline silicon solar cells

The techniques for production of multicrystalline or polycrystalline silicon are less critical, and hence cheaper, than those that required for single crystal material. The grain boundaries reduce the cell performance by blocking carrier flows, allowing extra energy levels in the forbidden gap, thereby providing effective recombination sites, and providing shunting paths for current flow across the p - n junction. To avoid significant recombination losses at grain boundaries, grain sizes in the order of at least a few millimeters are required. This also allows single grains to extend from the front to the back of a cell, providing less resistance to carrier flow and generally decreasing the length of grain boundaries per unit of cell. Such multicrystalline material is widely used for commercial solar cell production (Figure 3 - 3) .

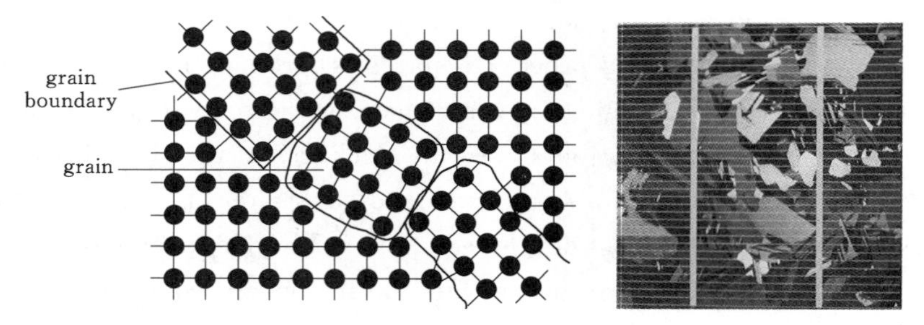

Figure 3 - 3 Crystal structure of the multicrystalline silicon material.

Polycrystalline ormulticrystalline silicon cells are made from cells cut from an ingot of melted and re - crystallized silicon. In the manufacturing process, molten silicon is cast into ingots of polycrystalline silicon, these ingots are then saw - cut into very thin wafers as per the mono - crystalline material and assembled into complete cells. Multicrystalline cells are cheaper to produce than mono - crystalline ones, due to the simpler manufacturing process. However, they tend to be slightly less efficient, with average efficiencies of around 22%. They have a speckled crystal reflective appearance, and again need to be mounted in a rigid frame, due to brittleness.

Specialized Vocabulary:

- shunting path 分流路径、分流电路
- recombination loss 复合损失
- grain size 晶粒尺寸
- ingot 铸块、锭
- molten silicon 熔融硅
- cast 浇铸
- speckled crystal reflective appearance 斑点晶体发光的外观
- rigid frame 钢架、刚性构架

3. 4 Amorphous silicon solar cells

Amorphous silicon solar cells are composed of silicon atoms in a thin homogenous layer rather than a crystal structure. Amorphous silicon absorbs light more effectively than crystalline silicon, so the cells can be thinner. For this reason, amorphous silicon is also known as a "thin film" PV technology. Amorphous silicon can be deposited on a wide range of substrates, both rigid and flexible, which makes it ideal for curved surfaces and "fold - away" modules. Amorphous cells are, however, less efficient than crystalline based cells, with typical maximum efficiencies of around 14%, but they are easier and therefore cheaper to produce. Their low cost makes them ideally suited for many applications where high efficiency is not required and low cost is important. One characteristic of amorphous silicon solar cells is that their power output reduces over time, particularly during the first few months, after which time they are basically stable. The quoted output of an amorphous panel should be produced after this stabilization (Figure 3 - 4).

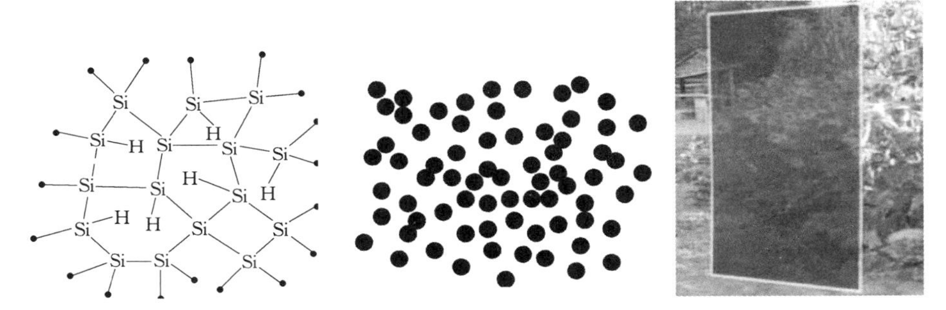

Figure 3 - 4 Crystal structure of the amorphous crystalline silicon material.

Amorphous silicon can be produced, in principle, even more cheaply than polycrystalline silicon. With amorphous silicon, there is no long - range order in the structural arrangement of the atoms, resulting in areas within the material containing unsatisfied, or "dangling" bonds. These in turn result in extra energy levels within the forbidden gap, making it impossible to dope the semiconductor when pure, or to obtain reasonable current flows in a solar cell configuration. It has been found that the incorporation of atomic hydrogen in amorphous silicon, to a level of 5%~10%, saturates the dangling bonds and improves the quality of the material. It also increases the bandgap (E_g) from 1.1eV in crystalline silicon to 1.7eV, making the material much more strongly absorbing for photons of energy above the latter threshold. The thickness of material required to form a functioning solar cell is therefore much smaller.

Specialized Vocabulary:

- homogenous 同质的
- homogenous layer 均质层、均匀层
- power output 输出功率
- long - range order 长程有序

- dangling' bonds 悬挂键、悬空键
- threshold 阈值
- plasma enhanced chemical vapor deposition (PECVD) 等离子体增强化学气相沉积
- atmospheric pressure chemical vapor deposition (APCVD) 常压化学气相沉积
- evaporation sputter 蒸发溅射

3.5 Ribbon silicon

The shortage of Si feedstock and the goal of reducingwatt peak (W) costs in PV is the driving force to look for alternatives to ingot grown multicrystalline Si wafers which have the higher share in the PV market. Ribbon Si was seen to be a candidate as no kerf losses occur, resulting in reduced Si costs per watt. In addition, there is no need for the energy consuming crystallization of the ingot and therefore energy payback times can be significantly reduced. However, the higher defect density in ribbon Si materials has to be taken into account during cell processing.

3.6 Czochralski silicon (Cz - Si)

For several decades, the terrestrial PV market has been dominated by p - type Czochralski silicon substrates. Continuous improvements in performance, yields and reliability have allowed an important cost reduction and the subsequent expansion of the PV market. Because of the lower cost of multicrystalline silicon (MC - Si) wafers, MC - Si cells emerged in the 1980s as an alternative to single - crystal wafers. However, their lower quality precluded the achievement of similar efficiencies to those of Cz, so that the figure of merit $/W has been quite similar for both technologies over a long time.

3.7 Silicon technology market

Silicon technology represents nowadays about 90% of the world photovoltaic market, and it is a well - established technology. Crystalline silicon (c - Si), Czochralski - material, and cast multicrystalline materials are the major contributors to the market, especially crystalline silicon (Figure 3 - 5) . Although the basic design of crystalline devices has remained essentially the same for 20 years, the vast growth of the photovoltaic market during the last 5 years is directly related to a major and dominant presence of more efficient, reliable and lower cost MC - Si modules.

Advantages of silicon technology are the good availability of the basic material for silicon production (quartz) with adequate physical properties for the preparation of photovol-

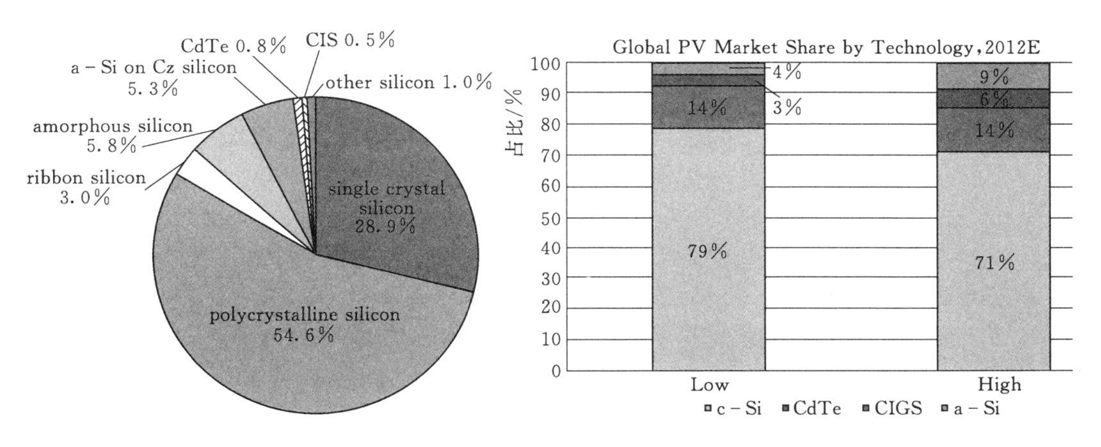

Figure 3-5 Distribution of cells' production by technology during 2002 and 2012.

taic devices, the relatively high energy conversion efficiency and the good stability of the cells when encapsulated, resulting in an expected lifetime of 30 years nowadays. Silicon technology challenges are related to the possible shortage of low-cost solar-grade silicon feedstock, the expense of crystal growth and wafer slicing, and the use of relatively thick (>200 micro) substrates to avoid breakage.

The basic solar cell manufacturing process diagram is shown in Figure 3-6.

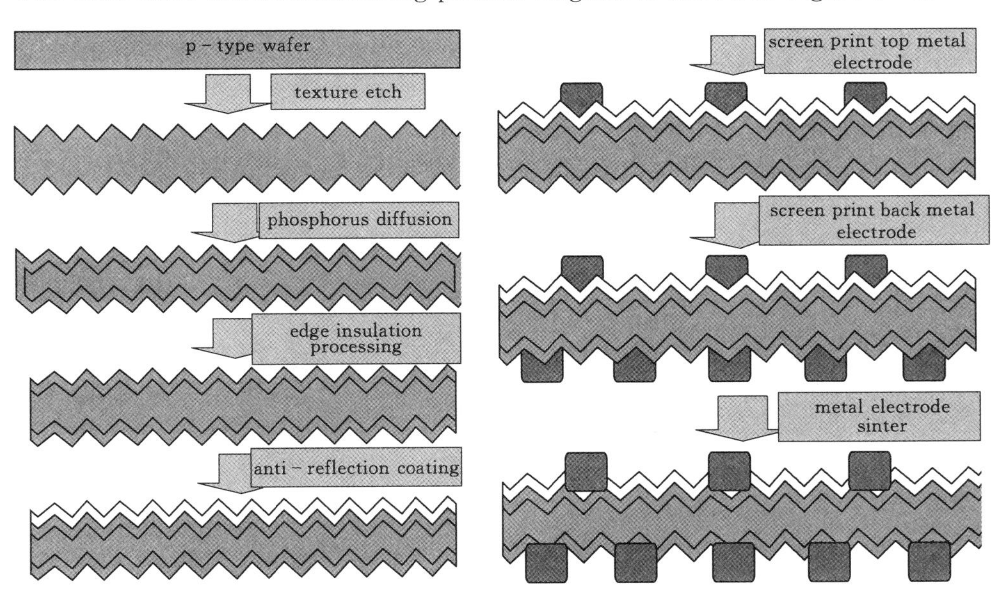

Figure 3-6 The basic solar cell manufacturing process diagram.

3.8 Maximum efficiencies for solar cell materials

The best efficiencies for all types of solar cells are shown in Figure 3-7 based on the

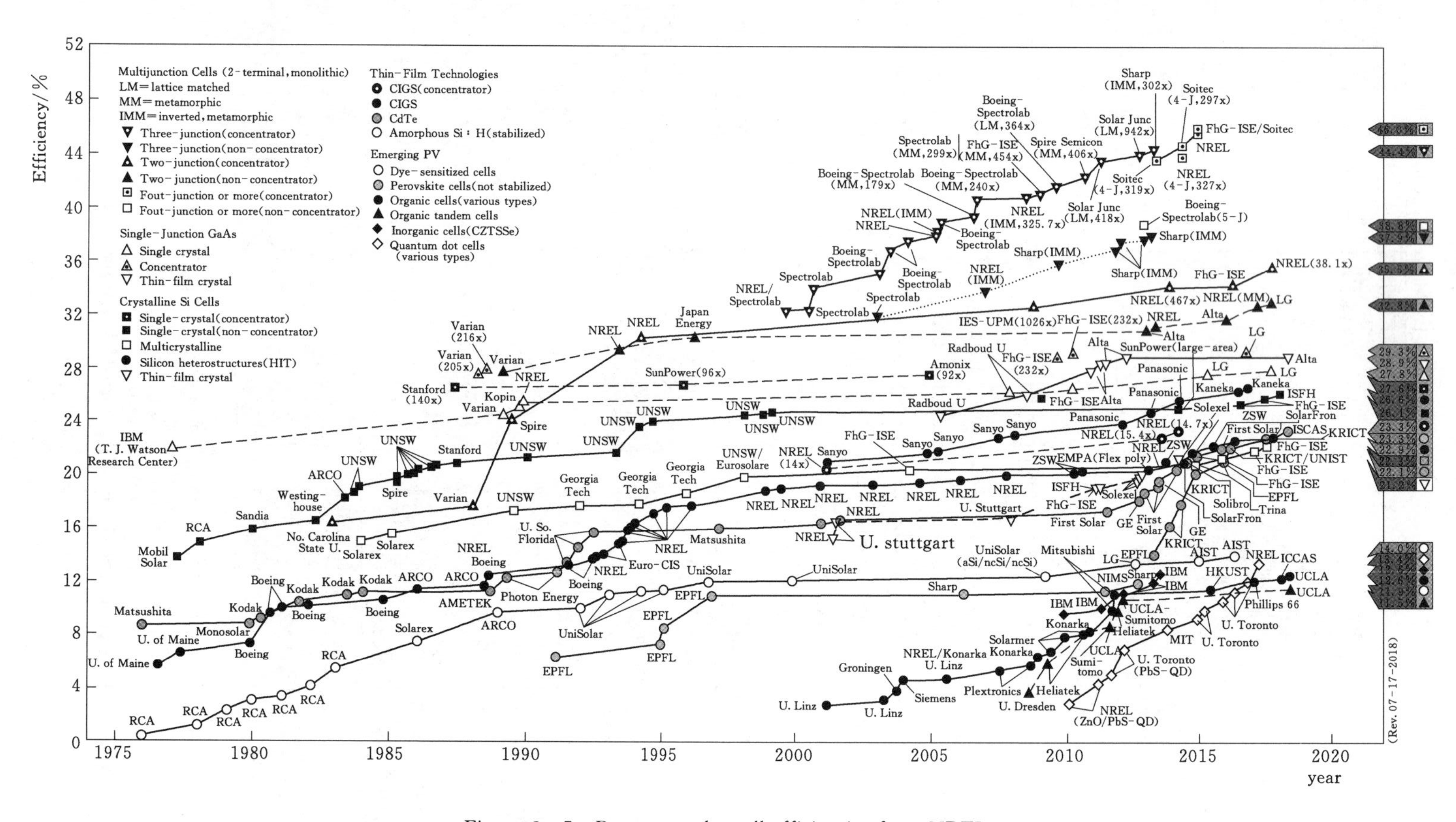

Figure 3-7 Best research-cell efficiencies from NREL.

(http://www.nrel.gov/pv/assets/pdfs/pv-efficiencies-07-17-2018.pdf.)

data from National Renewable Energy Laboratory (NREL) until 2017. The highest conversion efficiency is normally from concentrated solar cells of four or more junctions, which is 46%. For silicon based solar cells, the best efficiency is over 25% which is from monocrystalline silicon solar cells (non - concentrator) . The highest efficiency for compound solar cells is from single - junction GaAs with 28.8%. For the new generation solar cells, the highest efficiency is obtained by perovskite solar cell which is the latest developed cell and it is the fastest growing solar cell to date.

Specialized Vocabulary:

- depletion region 耗尽区/层，势垒区，阻挡层
- ribbon silicon 带硅
- Czochralski silicon (Cz - Si) 直拉单晶硅
- cast multicrystalline silicon 铸造多晶硅
- solar - grade silicon 太阳级硅
- wafer slicing 晶圆切片，硅片切割
- feedstock 原料
- National Renewable Energy Laboratory (NREL) 国家可再生能源实验室

3.9 Reading material (Please read the article and find out the specialized vocabulary)

The production of silicon solar cells[1]

The production of a typical silicon solar cell (Figure 3 - 8) starts with the carbothermic reduction of silicates in an electric arc furnace. In this process large amounts of electrical energy break the silicon - oxygen bond in SiO_2 via an endothermic reaction with carbon. Molten Si - metal with entrained impurities is withdrawn from the bottom of the furnace while CO_2 and fine SiO_2 particles escape with the flu - gas. A more detailed description of the process is given in the next section.

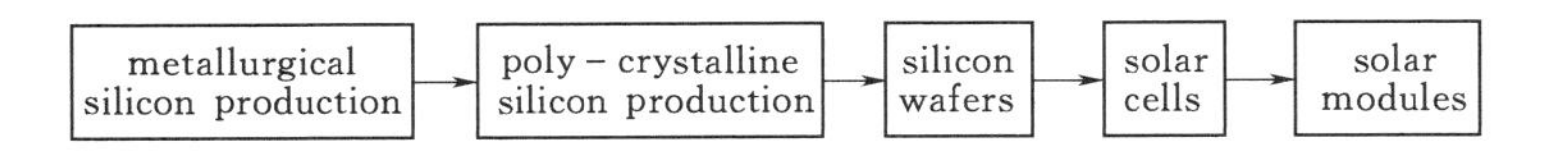

Figure 3 - 8 The supply chain for solar cell modules.

Metallurgical grade silicon (MG - Si) at about 98.5% purity is sold to many different markets. The majority of MG - Si is used for silicones and aluminum alloys. A much smaller

[1] Silicon solar cell production, computers and chemical engineering 2011, 35, 1439 - 1453.

portion is used for fumed silica, medical and cosmetic products and micro - electronics. A small but rapidly growing portion is used for solar applications. The price of MG - Si fluctuates with market conditions as seen in Figure 3 - 9.

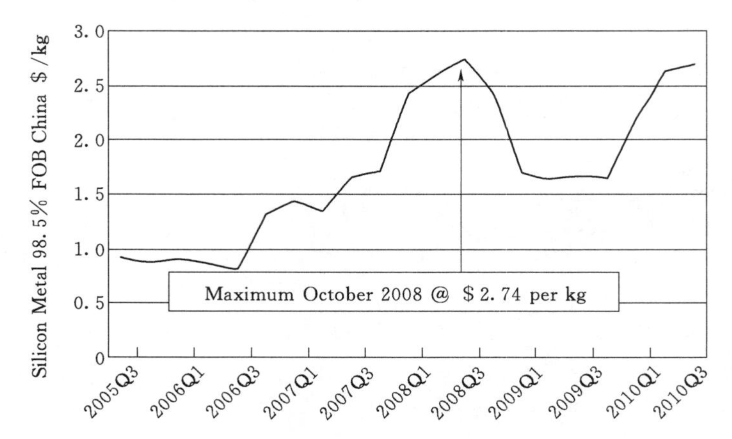

Figure 3 - 9 The price of silicon metal at 98.5% purity (MG - Si) FOB China.

Highly pure poly - silicon suitable for solar cells and microelectronics is typically produced in two steps. In the first step, MG - Si reacts with HCl to form a range of chlorosilanes, including tri - chlorosilane (TCS). TCS has a normal boiling point of 31.8℃ so that it can be purified by distillation. One process alternative for producing TCS is shown in Figure 3 - 10.

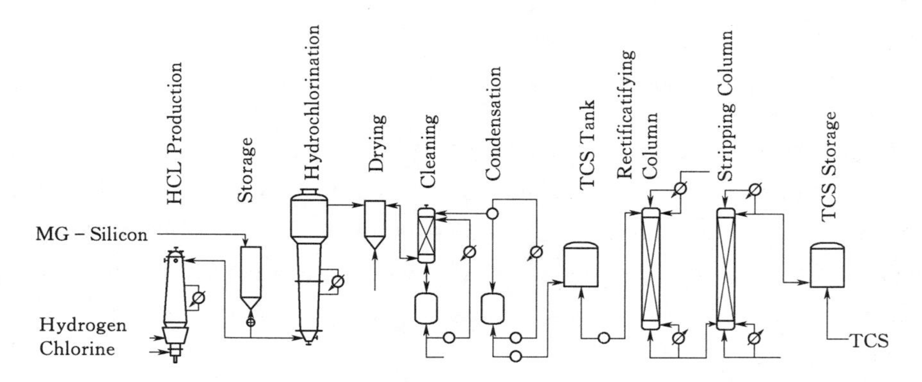

Figure 3 - 10 The production of highly pure TCS from MG - Si.

Poly - silicon is most often produced in the same manufacturing facility by pyrolysis of TCS in reactors that are commonly called Bell or Siemens reactors. In the Bell reactor, TCS passes over high purity silicon starter rods which are heated to about 1150℃ by electrical resistive heating. The gas decomposes as

$$2HSiCl_3 \longrightarrow Si + 2HCl + SiCl_4$$

Silicon deposits on the silicon rods as in a chemical vapor deposition process. 9N (99.999999999%) silicon is used for micro - electronics applications. Silicon which is 6N or better is called solar grade silicon (SOG - Si) and it can be used to produce high quality solar cells. The total process therefore takes silicon at low purity and converts it to poly - silicon at high purity. It is a very capital and energy intensive process, and many opportunities exist which the industry is currently investigating to reduce costs. A newfluidized bed process for making poly - silicon from silane is described in Section 4.

The annual price for solar grade silicon (Figure 3 - 11) went through a very sharp maximum in 2008 due to high demand and limited polysilicon production capacity. The increase in price was expected (Muller et al. , 2006; Woditsch & Koch, 2002) and led to a similar increase in the cost of wafers. The price then fell sharply due to the combined effect of capacity expansion and industry slow - down during the recent economic recession.

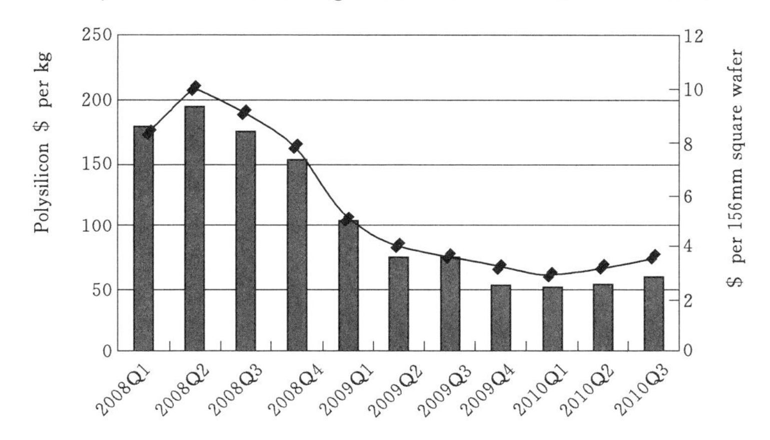

Figure 3 - 11 The correlation between the price of silicon wafers and polysilicon.

The price of solar grade silicon is expected to stabilize in the coming decade as new technologies are introduced and capacity is added to the supply chain. The classical TCS process was designed for micro - electronics manufacturing where the cost of silicon is not as critical as in the solar cell industry. Some companies have retrofitted their processes to produce solar rather than micro - electronics grade silicon. The pyrolysis process has been made suitable for high volume production of poly - silicon. Reactive separation and complex instead of simple distillation has been proposed to reduce energy requirements, and fluidized bed reactor technology is set to replace the Bell reactors during the next decade. Finally, progress has been made in making solar grade silicon directly using metallurgical routes (Braga, Moreira, Zampieri, Bachhin, & Mei, 2008); however, attempts have not been as successful as necessary for implementation. Nevertheless, it is very likely that solar grade silicon prices can be reduced to $ 25～30 per kg in the next decade if the tempo of industry expansion is maintained.

In the next step, it is necessary to produce thin wafers of silicon. They typically have a thickness of 200～350m and a resistivity of about 1/cm. The recent industry trend has been towards thinner wafers and new wafer saws are capable of producing wafers with a thickness of 180m. Thin wafers are advantageous since the material requirements are small, but they cannot be too thin since they are very brittle and difficult to process.

The most widely used wafering process consists of melting and re-solidifying pure silicon. These silicon ingots are then cut to form wafers as in the micro-electronics industry. The Czochralski (Cz) process produces large single crystalline silicon rods/ingots by slowly pulling out a rotating seed crystal from a molten bath. The review papers by Lan (2004) and Kakimoto et al. (2007) provide overviews of different methods that have been applied to simulate the static and dynamic behavior of Cz crystallization.

Optimization and control of the silicon crystallization process is challenging since the process integrates diverse physical phenomena (Brown, Wang, & Mori, 2001; Jana, Dost, Kumar, & Durst, 2006). The pulling rate and the rotation speed of the silicon boule as well as the cooling rate must be coordinated using control to maintain the quality of the crystals. Marangoni flows and buoyancy effects must be modeled (Vizman, Friedrich, & Mueller, 2007). The models must also capture crystal orientation, surface tension variation, radiation effects, impurity segregation and effects which impact fluid flow and heat transfer (Huang, Lee, Hsieh, Hsu, & Lan, 2004; Huh, Gosele, & Tan, 1995; Lan, 2004; Virzi, 1993). Multi-scale models can, in principle, address such problems but they are yet to be developed for the entire process (Habler et al., 2001). Current technology is capable of producing cylindrical rods up to 2m long and 30cm in diameter.

The Cz process is capital intensive and the energy requirements are high since the melt must be maintained at a high temperature for extended periods of time. It may take two days or more to produce a single silicon ingot.

The Bridgman process overcomes some of the problems associated with Cz crystallization by casting a silicon ingot in a quartz crucible. Batur, Srinivasan, Duval, and Singh (1995) developed control models for the process whereas Sonda, Yeckel, Daoutidis, and Derby (2005) studied the stability of the process using bifurcation theory. Kokh, Popov, Kokh, Krasin, and Nepomnyaschikh (2007) provideda study of how a rotating heat field in the Bridgman solidification process affects the melt flow whereas Miyazawa, Liu, and Kakimoto (2008) studied how the effect of the crucible affects the interfacial shape in the solidification process.

The Bridgman process provides a less costly alternative to Cz crystallization. However, there are disadvantages. The Bridgman process results in multi-crystalline rather than single-crystalline silicon. Multi-crystalline devices suffer from electron recombination at crystal boundaries which shunts the solar cell and reduces the overall conversion efficiency. Additionally, the crucible introduces impurities which further contribute to inferior device performance. A high quality, multi-crystalline cell has a conversion efficiency of 16% at present whereas the efficiency of the Cz produced cell is 18% and can be higher.

One major disadvantage of both the Cz and Bridgman processes is that the silicon blocks must be cut using wired saws to produce thin wafers. The wires used in the sawing process may be as thick as the wafers themselves and 50% or more of the material is therefore lost as dust or must be recycled atsignificant cost. Many processes have therefore been proposed to achieve wafering and crystallization in one continuous step. None of these have yet had a significant impact on industrial wafer production.

Solar cells are produced from silicon wafers in a sequence ofsteps. The wafers are first treated with chemicals to enhance optical and electrical properties. Silicon, a group 4 element, is doped with the neighboring group 3 and 5 elements, typically boron and phosphorous, to produce the p - n junction with associated surplus and deficiency of electrons in the conduction bands; a more thorough description of the effects of and reasons for doping is in Appendix A. Anti - reflection coating layers reduce reflection losses at the front surface by trapping incident light within the cell (Heine & Morf, 1995) . Front and back electrical contacts are added to complete the solar cell. Individual cells are finally integrated into panels and festooned with inverters and other systems to produce electricity matched to the end user's requirements.

Only 15 years ago it would be fair to characterize the solar cell industry as a cottage industry. It then began to grow at the breakneck pace of nearly 30% per year and is showing little sign of slowing down. Many consolidations are taking place and huge investments are being made around the globe, especially in China, where nearly 40% of all solar panels are produced. The economics of scale have started to play a more significant role as predicted by the KPMG Bureau for Economic Research and Policy Consulting (Solar energy, 1999) and prices are falling as industry experience grows.

Process optimization and advanced nonlinear control are needed in many areas. The entire supply chain needs to be managed and unit processes must be controlled to high accuracy. In the following sections, we describe three distinct problem areas where tools from process systems engineering have been used to provide new approaches for control, scale - up and conceptual design. Many more opportunities exist to apply systems engineering techniques to further optimize and control existing processes so that quality requirements are met and costs are kept as low as possible.

Exercises and Discussion

1. What are the differences of the structures among monocrystalline, multicrystalline and amorphous crystalline silicon materials?

2. What are the advantages and disadvantages of monocrystalline, multicrystalline and amorphous crystalline silicon?

3. What do you think of the future development of silicon - based solar cells?

Chapter 4

Inorganic Compounds Solar Cells

4.1 Cadmium telluride

Thin - film cadmium telluride (CdTe) solar cells are the basis of a significant technology with major commercial impact on solar energy production. Large - area monolithic thin - film modules demonstrate long - term stability, competitive performance, and the ability to attract production - scale capital investments.

During the past 20 years, there has been major progress in refining the basic CdTe cell structure. The highest current densities achieved are similar to crystalline GaAs when adjusted for small differences in band gap. Open - circuit voltage and fill factor are limited by high forward - current recombination and low carrier density, but have nevertheless achieved values about 80% as large as GaAs, again adjusted for band gap. There are some concerns about the diffusion of copper atoms, but significant degradation under normal operating conditions for well - fabricated cells is unlikely. Although cell - level basic research should certainly continue, the status of CdTe solar cells is clearly healthy enough to proceed with mainstream commercialization.

The future of CdTe thin - film photovoltaic devices in energy production and optical sensing is assured by the material properties, laboratory - scale device performance, and photovoltaic module implementation. Enhancing the viability of CdTe/CdS for terrestrial power generation depended on improving device performance, cost - effective translation of fabrication processes to the module scale, and cell and module stability, that have been achieved by First Solar. In addition to efficiency gains in single - junction devices, band gap tailoring of the absorber by alloying with other group Ⅱ B metals can facilitate development of multi - junction cells with efficiencies approaching 30%. Translating single - junction efficiency gains from batch processes to continuous module fabrication and developing monolithic multi - junction modules on a single superstrate can significantly reduce production costs. Achieving this goal required greater understanding of the relationship between processing conditions and critical material properties needed for high efficiency and good long - term stability. Structure of CdTe solar cell is shown in Figure 4 - 1.

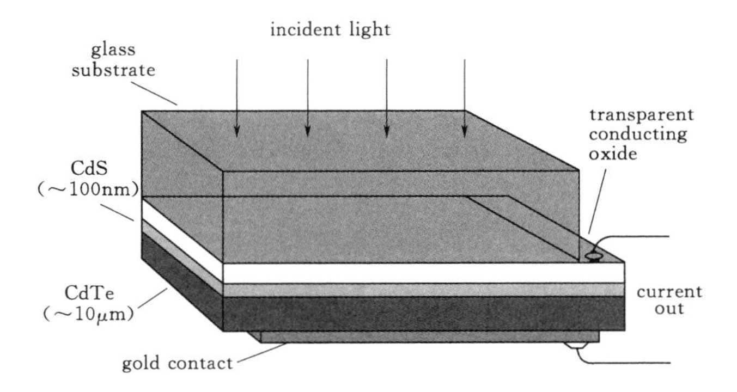

Figure 4-1 Structure of CdTe solar cell.

In addition to its basis for single-junction devices, CdTe can be alloyed with other ⅡB-ⅥA compounds to alter its band gap, allowing multi-junction cells to be designed. The multi-junction cell structures using CdTe-based wide band gap cells in monolithic structures must deal with the relationship between cell geometry and both processing temperature and chemical stability. Materials based on alloys between CdTe and other group ⅡB-ⅥA compounds allow a wide range of optoelectronic properties to be incorporated into devices by design. Semiconducting compounds of the form ⅡB-ⅥA provide a basis for the development of tunable materials, obtained by alloying different compounds in pseudobinary configurations. For photovoltaic hetero-junction devices, semiconductors using Cd, Zn, Hg cations and S, Se, Te anions exhibit a wide range of optical band gap, suggesting their potential for use in optimized device designed by tailoring material properties. The high optical absorption coefficients, 105/cm, and direct optical band gaps of many Ⅱ-Ⅵ semiconductors make them suitable for use in thin-film photovoltaic devices. For terrestrial photovoltaic applications, in which a band gap of 1.5eV is desired, considerable progress has been made in the development of solar cells based on the CdS-CdTe hetero-junction wherein $CdS_{1-y}Te_y$ and $CdTe_{1-x}S_x$ alloys have been shown to play a role in the device operation. For the development of next-generation, multi-junction cells, top cells with an absorber band gap of 1.7eV are required [12,13].

Specialized Vocabulary:

- cadmium telluride (CdTe) 碲化镉
- single-junction 单结
- multi-junction 多结
- hetero-junction 异质结
- current density 电流密度
- open-circuit voltage 开路电压
- fill factor 填充因子
- forward-current 正向电流

- carrier density 载流子密度
- soda lime glass 钠钙玻璃
- buffer layer 缓冲层
- close – space sublimation (CSS) 近空间升华法
- indium tin oxide (ITO) 铟锡氧化物
- optical sensing 光学传感
- terrestrial power generation 地面发电
- cost – effective 划算的、成本效益好的、合算的
- fabrication processes 制备工艺、合成过程
- single – junction devices 单结器件
- band gap tailoring 带隙调整
- multi – junction cells 多结电池
- processing conditions 加工条件、制造条件
- optoelectronic properties 光电性能
- heterojunction devices 异质结器件
- optical absorption coefficients 光吸收系数
- device operation 器件运行、器件工作、设备操作

4.2 Copper indium diselenide and related compounds

Cu (InGa) Se_2 – based solar cells have often been touted as being among the most promising of solar cell technologies for cost – effective power generation. This is partly due to the advantages of thin films for low – cost, high – rate semiconductor deposition over large areas using layers only a few microns thick and for fabrication of monolithically interconnected modules. Perhaps more importantly, high efficiencies have been demonstrated with Cu (InGa) Se_2 at both the cell and the module levels. Currently, the highest solar cell efficiency is 23.4% with 0.5cm^2 total area fabricated from the statistical efficiency of National Renewable Energy Laboratory (NREL) [14]. Furthermore, several companies have demonstrated large area modules with efficiencies above 12% including a confirmed 13.4% efficiency on a 3459cm^2 module by Showa Shell . Finally, Cu (InGa) Se_2 solar cells and modules have shown excellent long – term stability in outdoor testing if fully encapsulated. In addition to its potential advantages for large – area terrestrial applications, Cu (InGa) Se_2 solar cells have shown high radiation resistance, compared to crystalline silicon solar cells [17,18] and can be made very lightweight with flexible substrates, so they are also promising for space applications.

From its earliest development, $CuInSe_2$ was considered promising for solar cells because of its favorable electronic and optical properties including its direct band gap with high absorption coefficient and inherent p - type conductivity. As science and technology developed, it also became apparent that it is a very forgiving material since ① high efficiency devices can be made with a wide tolerance to variations in Cu (InGa) Se_2 composition [19,20]; ② grain boundaries are inherently passive so even films with grain sizes less than 1μm can be used; ③ device behavior is insensitive to defects at the junction caused by a lattice mismatch or impurities between the Cu (InGa) Se_2 and CdS. Structure of Cu (InGa) Se_2 solar cell is shown in Figure 4 - 2. However, there has always been concern about susceptibility to grain boundary degaration by moisture.

Figure 4 - 2 Structure of Cu (InGa) Se_2 solar cell.

The understanding of Cu (InGa) Se_2 thin films, as used in PV devices, is primarily based on studies of its base material, pure $CuInSe_2$. However, the material used for making solar cells is Cu (InGa) Se_2 containing significant amounts (of the order of 0.1%) of Na [24]. Even though the behavior of $CuInSe_2$ provides a good basis for the understanding of device - quality material, there are pronounced differences when Ga and Na are present in the films. More recently, Cu (InGa) Se_2 has been reviewed in the context of solar cells with an emphasis on electronic properties [25].

Deposition of CIGS thin films there can be by evaporation, sputtering and CVD - based technique, which are carried ont under high vacuum or low environment. There are also printing method using nanoparticles (inks) and liquid phase deposition method under atmosphere condition at 600～700℃.

Specialized Vocabulary:

- high - rate semiconductor deposition 高效/高速半导体沉积
- radiation resistance 耐辐射性、辐射抗性
- absorption coefficient 吸收系数
- lattice mismatch 晶格失配
- Aluminium doped zinc oxide (AZO) 掺铝氧化锌
- indium tin oxide (ITO) 铟锡氧化物
- Fluorine doped tin oxide (FTO) 掺氟氧化锡
- Chemical bath deposition 化学浴沉积法
- Sputtering 溅射

- Selenization 硒化法
- intermetallic 金属间化合物
- Liquid phase deposition 液相沉积法
- Radio frequency sputtering 射频溅射

4.3 GaAs

Gallium arsenide (GaAs) is a typical Ⅲ-Ⅴ compound semiconductor and has the similar Sphalerite crystal structure as silicon. GaAs hasa direct band gap, the band gap width is 1.42eV (300K) . GaAs has high light emitting efficiency and optical absorption coefficient, and it became the foundation material in the field of optoelectronic, playing a critical role in the solar cells field. GaAs is often used as a substrate material for the epitaxial growth of other Ⅲ-Ⅴ semiconductors including InGaAs and GaInNAs.

A very important application of GaAs is for high efficiency solar cells. GaAs is also known as a thin film single-crystalline device and are high-cost high-efficiency solar cells. In 1970, the first GaAs heterostructure solar cells were created by the team led by Zhores Alferov [26]. In the early 1980s, the efficiency of the best GaAs solar cells surpassed that of silicon solar cells, and in the 1990s GaAs solar cells took over from silicon as the cell type most commonly used for photovoltaic arrays for satellite applications. Later, dual- and triple-junction solar cells based on GaAs with germanium and indium gallium phosphide layers were developed as the basis of a triple-junction solar cell, which held a record efficiency of over 32% and can operate also with light as concentrated as 2000 suns. This kind of solar cell powers the rovers Spirit and Opportunity, which explore Mars' surface. Also many solar cars utilize GaAs in solar arrays. GaAs-based devices hold the world record for the highest-efficiency single-junction solar cell at 28.8%. This high efficiency is attributed to the extremely high quality GaAs epitaxial growth, surface passivation by the AlGaAs [28], and the promotion of photon recycling by the thin film design. Complex designs of $Al_xGa_{1-x}As$-GaAs devices can be sensitive to infrared radiation. Structure of GaAs solar cell is shown in Figure 4-3.

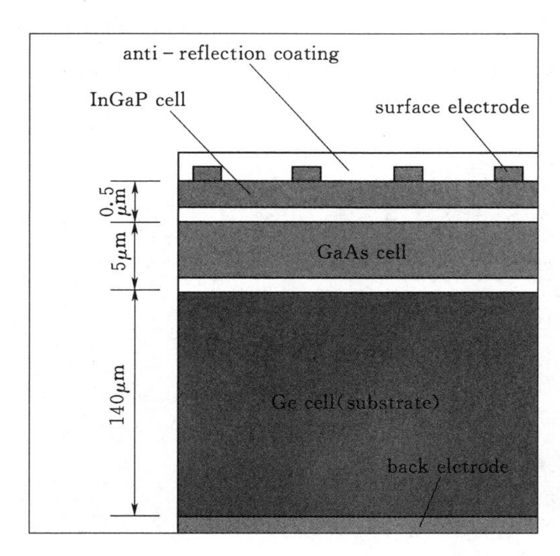

Figure 4-3 Structure of GaAs solar cell.

Specialized Vocabulary:

- epitaxial growth 外延生长
- heterostructure 异质结构，异质结
- light emitting efficiency 发光效率，出光效率
- foundation materials 基础材料
- Sphalerite crystal structure 闪锌矿晶体结构
- satellite applications 卫星应用
- dual - junction 双结
- triple - junction 三结
- Germanium 锗
- indium gallium phosphide 铟镓磷
- rovers 火星车
- infrared radiation 红外线照射

4.4 Reading material (Please read the article and find out the specialized vocabulary)

Inorganic photovoltaic cells

[From Inorganic photovoltaic cells, Robert W. Miles, Guillaume Zoppi, and Ian Forbes, materials today, 2007, 10 (11), 20 - 27]

Ⅲ - Ⅴ solar cells

Single junction Ⅲ - Ⅴ solar cells

The Ⅲ - Ⅴ compounds, such as GaAs, InP, and GaSb, have direct energy bandgaps, high optical absorption coefficients, and good values of minority carrier lifetimes and mobilities (in highly pure, single - crystalline material) . This makes them excellent materials for high - efficiency solar cells. The Ⅲ - Ⅴ materials used most widely for making single - junction solar cells are GaAs and InP, as both have near optimum energy bandgaps of 1.4eV. Originally, devices were formed by the diffusion of n - type dopants into wafers from single crystals produced using either the liquid - encapsulated Czochralski (LEC) method or a Bridgmann method. However, the highest conversion efficiencies confirmed under standard conditions are 25.8% for GaAs and 21.9% for InP single - junction cells, and were obtained with epitaxially grown homojunction structures produced in the US by the Kopin (Bedford, MA) and Spire (Westboro, MA) Corporations, respectively.

The disadvantage of using Ⅲ - Ⅴ compounds in photovoltaic devices is the very high cost of producing device - quality substrates or epitaxial layers of these compounds. Crystal imperfections, including unwanted impurities, severely reduce device efficiencies and alter-

native, lower cost deposition methods cannot be used. These materials are also easily cleaved and are significantly mechanically weaker than Si. The high density of the materials is also a disadvantage, in terms of weight, unless very thin cells can be produced to take advantage of their high absorption coefficients. These drawbacks have led to Ⅲ-Ⅴ compounds not being considered as promising materials for single-junction, terrestrial solar cells. The development of Ⅲ-Ⅴ based devices has been undertaken primarily because of their potential for space applications. There, the potential for high conversion efficiencies together with radiation resistance in the demanding environment of space power generation mitigates against the high materials cost. The high cost of cell manufacture can also be offset for terrestrial applications by using the high efficiency Ⅲ-Ⅴ cells in concentrator systems.

A recent development is that of quantum well (QW) cells made using GaAs and Ⅲ-Ⅴ alloys. A strain balanced, 50QW solar cell has a higher efficiency than a p-n-GaAs control cell. This device was grown by metal-organic chemical vapor deposition (MOCVD).

Ⅲ-Ⅴ multijunction devices

In a single-junction Si solar cell, 56% of the available energy is lost because photons with energies less than the bandgap are not absorbed, and photons with energies greater than the bandgap "thermalize", such that the excess energy over the bandgap is lost as heat. A range of studies have shown that using multijunction solar cells (sometimes referred to as tandem solar cells), such losses can be minimized leading to much higher efficiency devices. A landmark achievement demonstrating this concept was the development in 1990 of a GaAs/GaSb stacked cell with an efficiency greater than 30%. Much work since then has been on the development of stacked cells grown by MOCVD onto GaAs, InP, and more recently Ge substrates. The most efficient stacked cell devices are now produced by Spectrolab in Sylmar, California. A cross-sectional view of a device with an efficiency >39% is given in Figure 4-4.

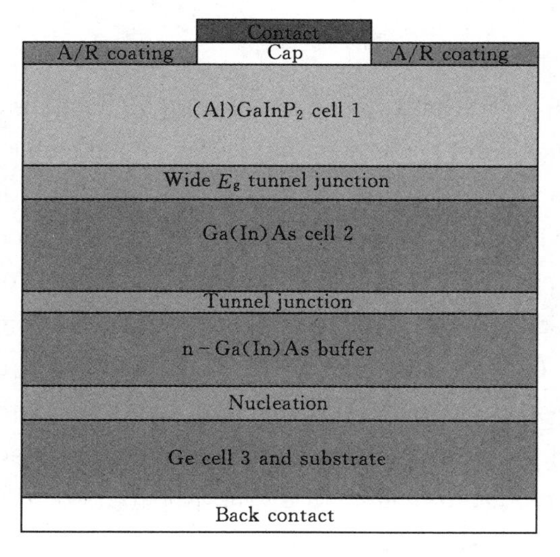

Figure 4-4 Cross-sectional view of a Ⅲ-Ⅴ multijunction device grown on a Ge substrate with 39.3% efficiency.

Thin-film solar cells based on compound semiconductors

Solar cells based on CdTe

With a direct optical energy bandgap of 1.5eV and high optical absorption coefficient for photons with energies greater than 1.5eV, only a few microns of CdTe are needed to absorb most of the incident light. Because only thin layers are needed, material costs are minimized, and because a short minority diffusion length (a few microns) is adequate, expensive materials processing can be avoided.

A cross - sectional view of a typical CdS/CdTe solar cell is shown in Figure 4 - 5 (a) .

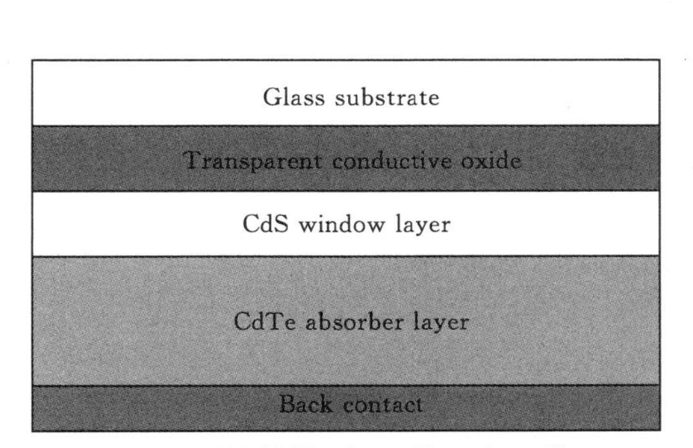

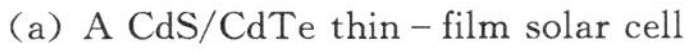

(a) A CdS/CdTe thin - film solar cell

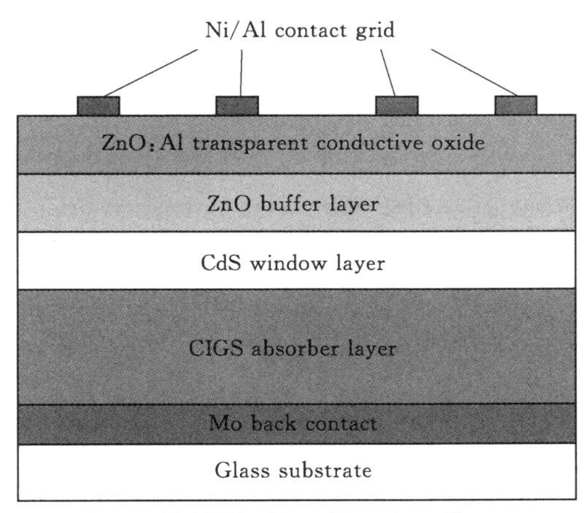

(b) A CdS/CIGS thin - flim solar cell

Figure 4 - 5 Cross - sectional views of thin - film solar cells based on the use of compound semiconductors.

A front contact is provided by depositing a transparent conductive oxide (TCO) onto the glass substrate. The TCO layer has a high optical transparency in the visible and near - infrared regions and high n - type conductivity. This is followed by the deposition of a CdS window layer, the CdTe absorber layer, and finally the back contact. For commercial devices, the CdS layer is usually deposited using either closed - space sublimation (CSS) or chemical bath deposition, although other methods have been used to investigate the fundamental properties of devices in the research laboratory. The CdTe p - type absorber layer, 3～10μm thick, can be deposited using a variety of techniques including physical vapor deposition (PVD), CSS, electrodeposition, and spray pyrolysis. To produce the most efficient devices, an activation process is required in the presence of $CdCl_2$ regardless of the deposition technique. This treatment is known to recrystallize the CdTe layer, passivate grain boundaries, and promote interdiffusion of the CdS and CdTe at the interface. Forming an ohmic contact to CdTe is difficult because the work function of CdTe is higher than all metals. This can be overcome by creating a thin p+ layer by etching the surface in bromine methanol or HNO_3/H_3PO_4 acid solution and depositing Cu - Au alloy or ZnTe: Cu. This creates a thin, highly doped region that carriers can tunnel through. However, Cu is a strong diffuser in CdTe and causes the performances to degrade with time. Another approach is to use a very low bandgap material, e. g. Sb_2Te_3, followed by Mo or W. This technique does not require a surface etch and the device performance does not degrade with time.

The most efficient CdTe/CdS solar cells (efficiencies up to 16. 5% 13) have been produced using a slightly different design to that shown in Figure 4 - 5 (a) . The improved efficiency is a result of the use of a Cd_2SnO_4 TCO layer which is more transmissive and conductive than the classical SnO_2-based TCOs, and the inclusion of a Zn_2SnO_4 buffer layer which improves the quality of the device interface. Two companies currently manufacture CdTe-based modules: First Solar and Antec Solar. Both use thermal sublimation processes and have managed to produce modules with baseline efficiencies of 89%. First Solar current production costs are \$1.25/Wp for a 99MWp/year manufacturing line, and projected output is 275MWp/year for 2008 with estimated cost below the \$1/Wp barrier. In August 2007, CdTe modules manufactured by First Solar were installed on the roof of a logistic centre in Ramstein, Germany. The installation is capable of generating 2.5MWp of power, i.e. it is currently the largest thin-film BIPV generation project using thin-film solar cells.

Solar cells based on chalcopyrite compounds

The first chalcopyrite solar cells developed were based on the use of $CuInSe_2$ (CIS). It was, however, rapidly realized that incorporating Ga to produce the solid solution Cu (In, Ga) Se_2 (CIGS), results in a widening of the energy bandgap to 1.3eV and an improvement in material quality, resulting in solar cells with enhanced efficiencies. CIGS has a direct energy bandgap and high optical absorption coefficient for photons with energies greater than the bandgap, such that only a few microns of material are needed to absorb most of the incident light, with consequent reductions in material and production costs. The best CIGS solar cells are grown on sodalime glass in the sequence: back contact, absorber layer, window layer, buffer layer, TCO, and then the top contact grid. A typical structure is shown in Figure 4-5 (b). CIGS solar cells have been produced with efficiencies of 19.5% and modules with efficiencies of 13.4%. The back contact is a thin film of Mo deposited by magnetron sputtering, typically 500~1000nm thick. The CIGS absorber layer is formed mainly by (i) the coevaporation of the elements either uniformly deposited or using the so-called three-stage process, or (ii) the deposition of the metallic precursor layers followed by selenization and/or sulfidization. Coevaporation yields devices with the highest performance while the latter deposition process is preferred for large-scale production. Both techniques require a processing temperature >500℃ to enhance grain growth and recrystallization. Another requirement is the presence of Na, either directly from the glass substrate or introduced chemically by evaporation of a Na compound. The primary effects of Na introduction are grain growth, passivation of grain boundaries, and a decrease in the absorber layer resistivity. This usually yields an increase of 12% in efficiency for Na concentrations <1%. The junction is usually formed by the chemical bath deposition of a thin (50~80nm) window layer. CdS has been found to be the best material, but alternatives such as ZnS, ZnSe, In_2S_3, (Zn, In) Se, Zn (O, S), and MgZnO can also be used. The buffer layer can be deposited by chemical bath deposition, sputtering, chemical vapor deposition, or evaporation, but the highest efficiencies have been achieved when using a wet process as a result of the presence of Cd^{2+} ions. A 50nm intrinsic ZnO buffer layer is then deposited and acts as to prevent any shunts. The TCO layer is usually ZnO： Al 0.5～1.5μm. The cell is finally completed by depositing a metal grid contact Ni/Al for current collection. The main CIGS manufacturers are Würth Solar, Avancis (formerly Shell Solar), and Global Solar. Numerous other ventures are engaged worldwide in the development of CIGS - based photovoltaic products.

Exercises and Discussion

1. Please describe the structure and working mechanism of CdTe, CIGS and GaAs solar cells, respectively.

2. What's the difference between CdTe, CIGS and GaAs solar cells?

Chapter 5

Concentrator Solar Cells

5.1 Concentrator solar cells

Photovoltaic concentrators are based on the principle of the use of a lens or mirror to collect the sun's power onto a small photovoltaic device in order to increase the energy conversion efficiency with the same cell active area and substitute the expense of a larger solar cell by a cheaper optical system (Figure 5-1). This allows for a reduction in the cell area required for producing a given amount of power. The goal is to significantly reduce the cost of electricity generated by replacing expensive PV converter area with less expensive optical material. This approach also provides the opportunity to use higher performance PV cells that would be prohibitively expensive without concentration. As a result, concentrator modules can easily exceed 30% energy conversion efficiency. In the future, the use of multijunction cells is expected to increase this to over 40%. While the concept is simple, and has been examined since the time of the earliest interest in terrestrial photovoltaics, the practice has proven to be deceptively difficult.

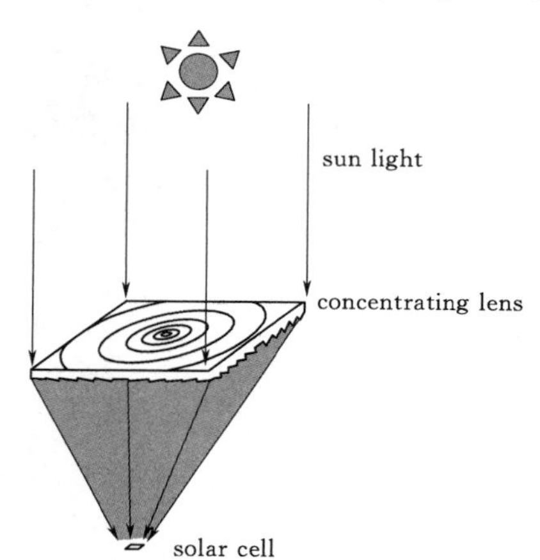

Figure 5-1 Concentrator solar cells.

Concentrator research has focused much effort on the PV cells themselves, which are now well developed and available commercially. The main remaining technical barriers, however, are due to the difficult cell packaging requirements stemming from the high heat flux and electrical current density, plus the need for more cost-effective and reliable tracking systems and module designs. The main market barriers have been due to the fact that concentrating systems, which in most cases must track the sun, are not well suited to the existing PV market that serves small remote loads and, more recently, building integrated applications. Another disadvantage of concentrator systems is the restriction to locations with a high proportion of direct radiation and they require a sun - tracking system as the aperture is small. A successful market implementation of PV - concentrators has not yet been reached.

The concentration ratio (ratio of module aperture area to cell area) varies from 2 to 4 in static concentrator designs that require no sun tracking to over 1000 times in some two - axis tracking systems. The means of optical concentration includes a variety of two - axis and one - axis reflective and refractive approaches, as well as many novel means such as luminescent and holographic concentrators. The development of concentrators has been aided and impacted by the parallel development of materials and other technologies. For example, the once cumbersome aspect of finding and tracking the sun is now made relatively straightforward by the emergence of very low - cost computing technology and the Global Positioning System. On another path, developments in the global semiconductor industry often have direct application to concentrator cells. Examples include larger wafers, improved processing equipment, the emergence of organo - metallic chemical vapor deposition (OMCVD) for fabricating multijunction Ⅲ-Ⅴ cells, and improved packaging materials with superior thermal properties. Many of the technical issues facing further concentrator development can be thought of as material issues. These include the development of polymer reflectors with improved weather resistance and lower cost molding methods for Fresnel lenses. In other words, concentrator development takes place in the larger technology arena. New material and technology developments may come from any direction and make possible what was only a dream previously. Unfortunately, tracking the sun is still effected by the distinctly nineteenth - century technology of gears and motors. The necessity for tracking remains concentrating PV's Achilles heel.

A key issue for PV concentrators is the necessity to develop energy - efficient and cost - effective solar cells that can operate at high sunlight flux (>500 suns) and the availability of economical concentration systems. In addition, a market for PV - concentrators still has to be developed. Concentrators were conceived as a vehicle to generate large amounts of nonpolluting renewable energy. As yet, costs are still too high to compete with fossil fuel - fired generation, or even the most direct renewable competitors - wind power and direct solar. The cost gap is narrowing, however, and there appears a likelihood that in the future concentrator systems will find cost - effective applications.

Specialized Vocabulary:

- concentrator solar cell 聚光太阳电池
- concentration ratio 聚光比，聚光比例
- Fresnel lenses 菲涅尔透镜
- photovoltaic concentrators 光伏聚光器
- lenses 透镜
- mirrors 反射镜、反光镜

- optical lens 光学透镜
- active area 有效面积
- technical barriers 技术瓶颈、技术壁垒
- heat flux 热通量、热流
- cell packaging 电池封装、电池包装
- sun-tracking system 太阳跟踪系统
- two-axis tracking systems 双轴跟踪系统
- one-axis 单轴
- reflective 反射，反光
- refractive 折射，折射的
- luminescent 发光，发光的，冷光的，场致发光
- holographic concentrators 全息聚光、全息聚光器
- global positioning system (GPS) 全球定位系统
- organo-metallic chemical vapor deposition (OMCVD) 有机金属化学气相沉积
- polymer reflectors 聚合物反射镜
- molding methods 成型方法
- Fresnel lenses 菲涅尔透镜
- sunlight flux 阳光通量
- fossil fuel-fired generation 化石燃料发电

5.2 Reading material (Please read the article and find out the specialized vocabulary)

5.2.1 Tandem solar cells (The solar spark, http://www.thesolarspark.co.uk/the-science/solar-power/tandem-solar-cells/)

What arethe tandem solar cells?

Tandem cells are effectively a stack of different solar cells on top of each other. By arranging them like this, we can capture more energy from the sun. If, for example, a solar cell is designed to work really well when it absorbs blue light, we could put it next to one that absorbs green light well and one that absorbs red light well so that we can capture more energy from the sun.

If you think about how TV and computer screens are designed, they use a combination of red, blue and green light to be able to make all the different colors you would need. By using all three of them, you can make a white light. In reverse, if we want to collect all the visible white light we want to collect all three colors blue, red and green.

Why use tandem solar cells?

There is a limit to the maximum efficiency that some types of solar cells, known as "single - junction cells" (e. g. silicon), can ever reach called the "Shockley - Queisser limit". If we look closely at sunlight, we can see that it is made up of photons (particles of light) that have a range of wavelengths, and therefore a range of energies. This is why we sometimes see rainbows when it rains; the water splits up the sunlight into its different colors, and each color has a different wavelength. There are also some parts of the solar spectrum (Figure 5 - 2) that we can't see with the naked eye - infrared and ultraviolet light.

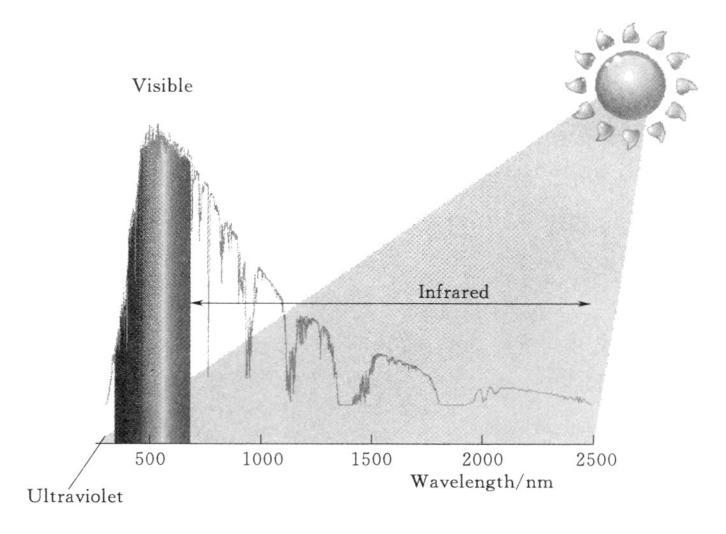

Figure 5 - 2 Solar spectrum.

Semiconductors, one of the components used in solar cells, have what is known as a "band gap", which is basically an energy range which needs to be overcome before the semiconductor will conduct electricity. The size of the band gap determines both the energy of light (and proportion of that light) that can be absorbed as well as the maximum power you can get from the cell. The power of a solar cell is determined by the current (related to the number of electrons in the conduction band) and the voltage (related to the size of the band gap) (Figure 5 - 3). For maximum power, we want both a large current and a large voltage. A large voltage is achieved by having a large band gap, but the bigger the band gap, the higher the amount of energy the photons of light need to have to be absorbed by the semiconductor. If the photons of light don't have enough energy, electrons will not be able to jump to the conduction band and we'll get a small current.

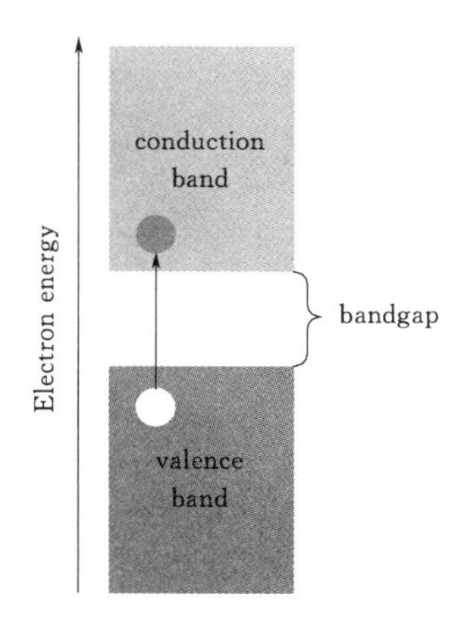

Figure 5 - 3 The relationship between bandgap and electron energy.

So can we just choose a semiconductor with a small band gap? Unfortunately not, for a couple of reasons. Just as photons with too *little* energy won't result in an energy output from our solar cell, photons with too *much* energy aren't so good either! The extra energy they have is just lost. Also, since the size of our band gap determines the voltage we get out of our cell, we don't want one which is very small, since this will mean a low voltage and therefore low power output.

It's a trade off between the two effects; if you change the band gap size to increase the voltage, the current will decrease, and vice versa. It has been worked out that the best balance settles on a theoretical maximum efficiency of around 33%.

Scientists are constantly trying to develop clever ways to overcome this limit, and have already come up with a few possible solutions. Tandem cell solar cells are one example. By using different solar cells which absorb different parts of the visible light spectrum, the value of that theoretical limit can be increased. Other types of solar cells which could do this include Solar Concentrators and Excitonic Solar Cells which use quantum dots.

Stacking the cells

The order of the cells when they are stacked together is important. Blue light has more energy than green or red so the semiconductor material that absorbs it has a bigger band gap (Figure 5 - 4) .

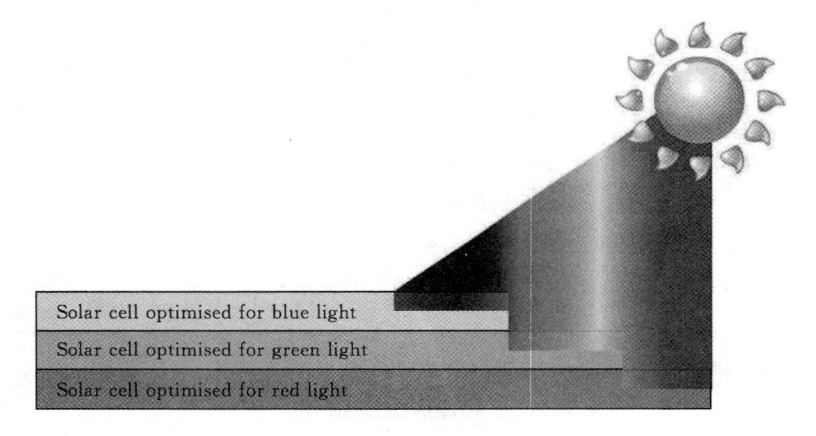

Figure 5 - 4 The order of the cells which stacked together is crucial to the absorption light.

The semiconductor material that absorbs red light well has a smaller band gap. This material can also absorb blue light but it is a waste of energy to do so. It is like building a bridge over a tiny stream that you could just step over, a waste of time and effort.

So to make use of the sunlight most efficiently, the light should reach the blue light absorber first so that this energy can be removed. However, the green and red light will not have enough energy to jump this band gap so they pass straight through to the next layer. Here the green light, the next highest in energy, can be absorbed and the red light can pass on through to a third layer. By doing this, you can get the most out of the sunlight.

Parallel and series circuits

You can also connect the different solar cell layers up in various ways. If you connect them in a parallel circuit you can get more current (rate of electron flow) from the system, and if you connect them into a series circuit you can get more voltage (potential or "push") from the system. How you connect them may depend on what you want to use your solar cells for.

5. 2. 2 Concentrating Photovoltaics (CPV) (From Green Rhino Energy)

Principle

In Concentrating Photovoltaics (CPV), a large area of sunlight is focused onto the solar cell with the help of an optical device. By concentrating sunlight onto a small area, this technology has three competitive

advantages:

• Requires less photovoltaic material to capture the same sunlight as non - concentrating PV.

• Makes the use of high - efficiency but expensive multi - junction cells economically viable due to smaller space requirements.

• The optical system comprises standard materials, manufactured in proven processes. Thus, it is less dependant on the immature silicon supply chain. Moreover, optics are less expensive than cells.

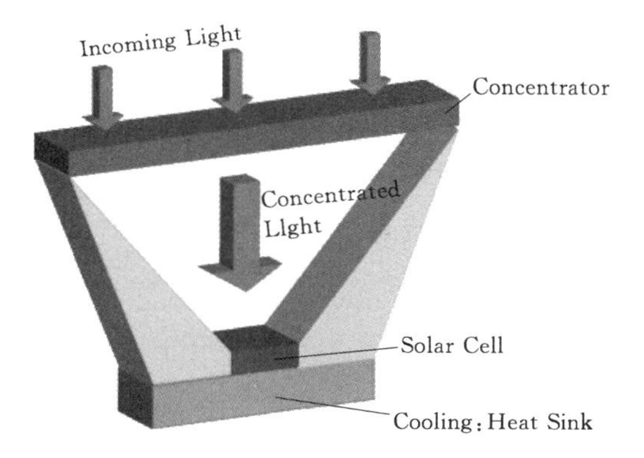

Concentrating light, however, requires direct sunlight rather than diffuse light, limiting this technology to clear, sunny locations. It also means that, in most instances, tracking is required.

Despite having been researched since the 1970s, it has only now entered the solar electricity sector as a viable alternative. Being a young technology, there is no single dominant design.

The most common classification of CPV - modules is by the degree of concentration, which is expressed in number of "suns" . E. g. "3x" means that the intensity of the light that hits the photovoltaic material is 3 times than it would be without concentration.

	Low concentration	Medium concentration	High concentration
Degree of concentration	2～10	10～100	>100
Tracking	No tracking necessary	1 – axis tracking sufficient	Dual axis tracking required
Cooling	No cooling required	Passive cooling sufficient	Active cooling reuqired in most instances
Photovoltaic Material	High – quality silicon		Multi – junction cells

Concentration

Here are some examples of concentrator technologies and examples for both line and point concentrators. Although there might be differences in execution or materials used, most designs will follow one of those concepts.

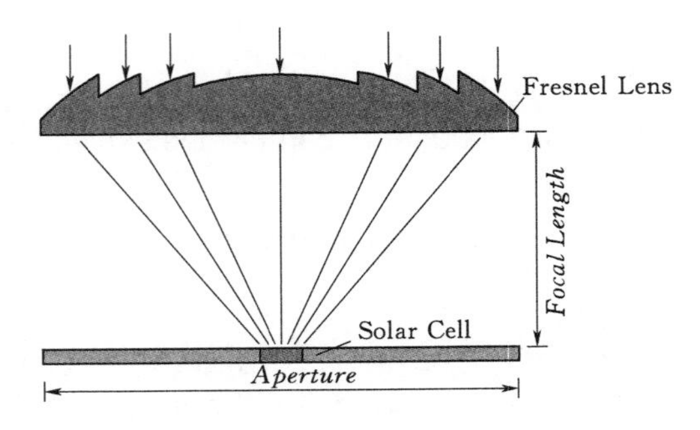

Fresnel Lens

A Fresnel lens, named after the French physicist, comprises several sections with different angles, thus reducing weight and thickness in comparison to a standard lens. With a Fresnel lens, it is possible to achieve short focal lenght and large aperture while keeping the lens leight.

Fresnel lenses can be constructed

- in a shape of a circle to provide a point focus with concentration ratios of around 500, or in cylindrical shape to provide line focus with lower concentration ratios.

With the high concentration ratio in a Fresnel point lens, it is possible to use a multi – junction photovoltaic cell with maximum efficiency. In a line concentrator, it is more common to use high efficiency silicon.

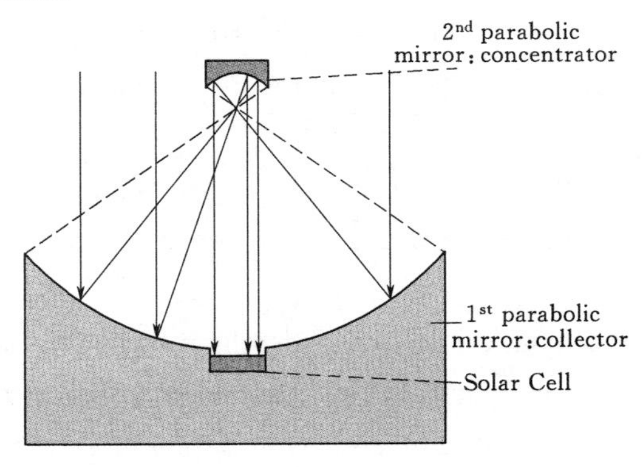

Parabolic Mirrors

Here, all incoming parallel light is reflected by the collector (the first mirror) through a focal point onto a second mirror. This second mirror, which is much smaller, is also a parabolic mirror with the same focal point. It reflects the light beams to the middle of the first parabolic mirror where it hits the solar cell.

The advantage of this configuration is that it does not require any optical lenses. However, losses will occur in both mirrors. SolFocus has achieved a concentration ratio of 500 in point concentrator - shape with dual axis - tracking.

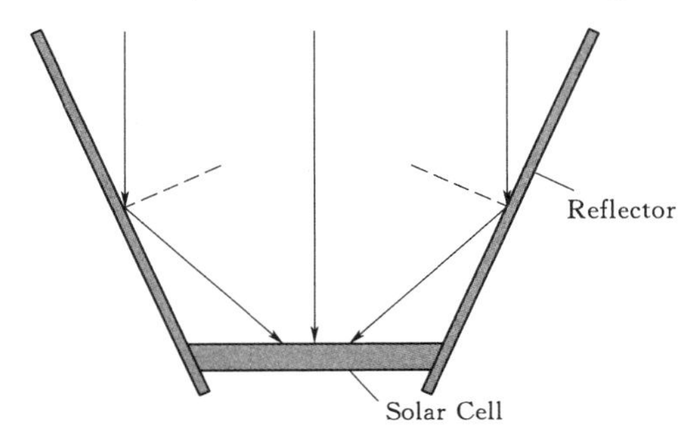

Reflectors

Low concentration photovoltaic modules use mirrors to concentrate sunlight onto a solar cell. Often, these mirrors are manufactured with silicone - covered metal. This technique lowers the reflection losses by effectively providing a second internal mirror. The angle of the mirrors depends on the inclination angle and latitude as well as the module design, but is typically fixed. The concentration ratios achieved range from 1.5～2.5. Low concentration cells are usually made from monocrystalline silicon. No cooling is required. The largest low - concentration photovoltaic plant in the world is Sevilla PV with modules from three companies: Artesa, Isofoton and Solartec.

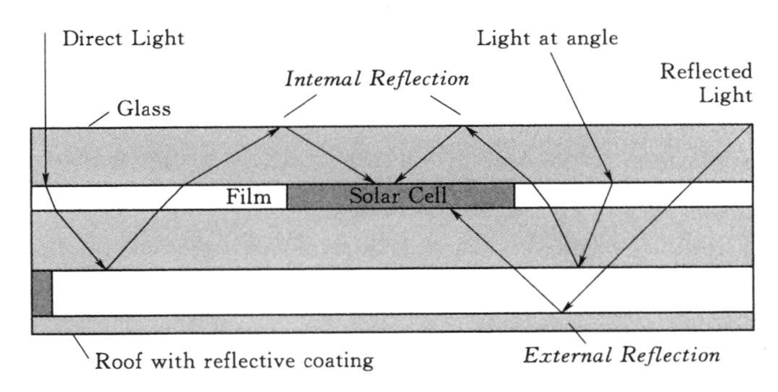

Luminescent Concentrators

In a luminescent concentrator, light is refracted in a luminescent film, and then being channeled towards the photovoltaic material. This is a very promising technology, as it does not require optical lenses or mirrors. Moreover, it also works with diffuse light and hence does not need tracking. The concentration factor is around 3.

There are various developments going on. For instance, Covalent are using an organic material for the film, whilst Prism Solar use holographic film.

Furthermore, this concentrator does not need any cooling, as the film could be constructed such that wavelengths that can not be converted by the solar cell would just pass thru. Hence, unwanted wavelengths would be removed.

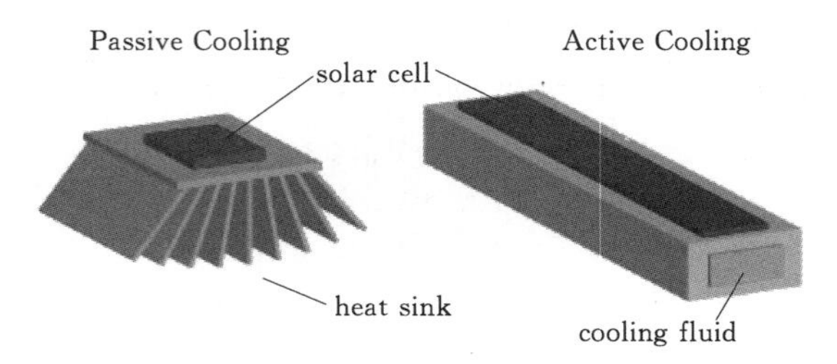

Cooling

Most concentrating PV systems require cooling.

Passive Cooling: Here, the cell is placed on a cladded cermaic substrate with high thermal conductivity. The ceramic also provides electrical isolation.

Active Cooling: Typically, liquid metal is used as a cooling fluid, capable of cooling from 1700℃ to 100℃.

Examples

Low - concentration modules

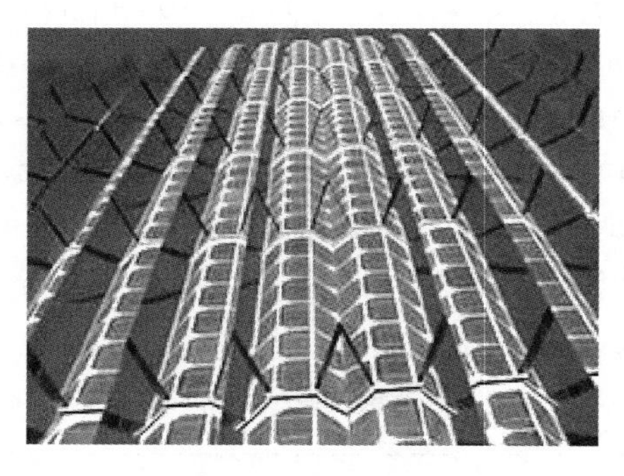

Low - concentration PV modules using mirrors without further tracking.

Linear Fresnel concentrator devices by Entech Solar.

High - concentration modules

High concentration 300x "Diamond Power" series byEnFocus has been specifically developed for rooftop installations, including dual - axis tracking.

Parabolic mirrors achieve 500x concentration in devices developed by SolFocus.

Exercises and Discussion

1. Please cite some examples of concentrator solar cells.
2. Which is the important component in concentrator solar cells? Why?

Chapter 6

Organic Solar Cells, DSSCs, PSCs and QDSCs

6.1 Organic solar cells

Organic solar cells (Figure 6 – 1) can be colored, transparent, and applied to flexible, lightweight films. They generate electricity even under cloudy sky. A type of polymer solar cell that uses organic electronics, a branch of electronics that deals with conductive organic polymers or small organic molecules, for light absorption and charge transport to produce electricity from sunlight by the photovoltaic effect.

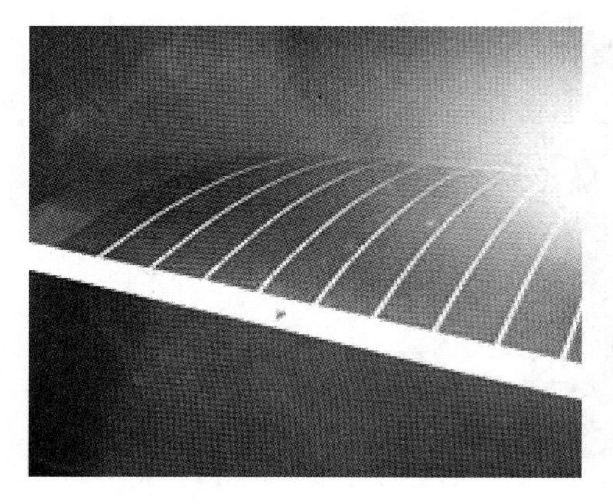

Figure 6 – 1 Organic solar cell.

Most of the plastic used in organic solar cells has low production costs in high volumes. Combined with the flexibility of organic molecules, organic solar cells are potentially cost – effective for certain photovoltaic applications. Molecular engineering (e. g. changing the length and functional group of polymers) can change the energy gap, which allows chemical change in these materials. The optical absorption coefficient of organic molecules is high, so a large amount of light can be absorbed with a small amount of materials. The main disadvantages associated with organic photovoltaic cells are low efficiency, low stability and low strength compared to rigid inorganic photovoltaic cells.

A photovoltaic cell is a specialized semiconductor diode that converts visible light into direct current (DC) electricity. Some photovoltaic cells can also convert infrared (IR) or ultraviolet (UV) radiation into DC. A common characteristic of both the small molecules and polymers used in photovoltaics is that they all have large conjugated systems. A conjugated system is formed where carbon atoms covalently bond with alternating single and double bonds; in other words these are chemical reactions of hydrocarbons. The pz electron orbitals of these hydrocarbons delocalize and form a delocalized bonding π orbital with a π^* antibonding orbital. The delocalized π orbital is the highest occupied molecular orbital

(HOMO), and the π^* orbital is the lowest unoccupied molecular orbital (LUMO). The separation between HOMO and LUMO is considered the band gap of organic electronic materials. The band gap is typically in the range of 1～4eV [31]. When these materials absorb a photon, an excited state is created and confined to a molecule or a region of a polymer chain. The excited state can be regarded as an electron-hole pair bound together by electrostatic interactions, i. e. excitons. In photovoltaic cells, excitons are broken up into free electron-hole pairs by effective fields. The effective fields are set up by creating a heterojunction between two dissimilar materials. Effective fields break up excitons by causing the electron to fall from the conduction band of the absorber to the conduction band of the acceptor molecule. It is necessary that the acceptor material has a conduction band edge that is lower than that of the absorber material [32].

There are different types of junctions for organic PV cells. Among them, single layer organic photovoltaic cells are the simplest of the various forms of organic photovoltaic cells. These cells are made by sandwiching a layer of organic electronic materials between two metallic conductors, typically a layer of indium tin oxide (ITO) with high work function and a layer of low work function metal such as Al, Mg or Ca. The basic structure of such a cell is illustrated in Figure 6-2 (a). In practice, single layer organic photovoltaic cells of this type do not work well. They have low quantum efficiencies ($<1\%$) and low power conversion efficiencies ($<0.1\%$). A major problem with them is that the electric field resulting from the difference between the two conductive electrodes is seldom sufficient to break up the photo-generated excitons. Often the electrons recombine with the holes rather than reach the electrode. To deal with this problem, the multilayer organic photovoltaic cells were developed. This type of organic photovoltaic cell contains two different layers in between the conductive electrodes [Figure 6-2 (b)]. These two layers of materials have differences in electron affinity and ionization energy. Therefore, electrostatic forces are generated at the interface between the two layers. The materials are chosen to make the differences large enough, so these local electric fields are strong, which may break up the excitons much more efficiently than the single layer photovoltaic cells do. The layer with higher electron affinity and ionization potential is the electron acceptor, and the other layer is the electron donor. This structure is also called a planar donor-acceptor heterojunction [32]. The diffusion length of excitons in organic electronic materials is typically of the order of 10nm. In order for most excitons to diffuse to the interface of layers and break up into carriers, the layer thickness should also be in the same range as the diffusion length. However, a polymer layer typically needs a thickness of at least 100nm to absorb enough light. At such a large thickness, only a small fraction of the excitons can reach the heterojunction interface.

To address this problem, a new type of heterojunction photovoltaic cells has been designed, which is the dispersed heterojunction photovoltaic cells. In this type of photovoltaic cell, the electron donor and acceptor are mixed together, forming a polymer blend [Figure 6-2 (c)]. If the length scale of the blend is similar to the exciton diffusion

length, most of the excitons generated in either material may reach the interface, where excitons break efficiently. Electrons move to the acceptor domains then are carried through the device and collected by one electrode, and holes are pulled in the opposite direction and collected at the other side [33].

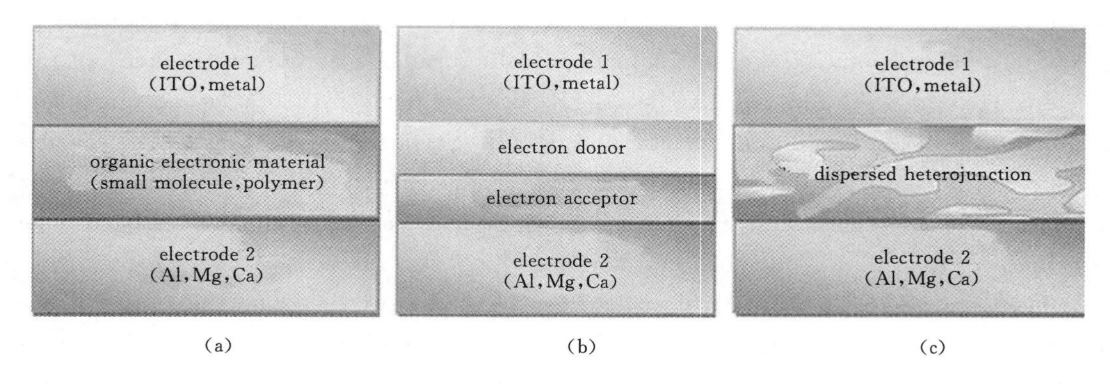

Figure 6 - 2 Sketch of a single layer, a multilayer and a dispersed junction photovoltaic cell.
(Source: Organic photovoltaic films, Current opinion in solid state and materials science, 2002, 6, 87.)

Difficulties associated with organic photovoltaic cells include their low external quantum efficiency (up to 70%)[34] in comparison with inorganic photovoltaic devices; due largely to the large band gap of organic materials. Instabilities against oxidation and reduction, and recrystallization are issues. Temperature variations can also lead to device degradation and decreased performance over time. This occurs to different extents for devices with different compositions, and is an area into which active research is taking place [35]. Other important factors include the exciton diffusion length; charge separation and charge collection; and charge transport and mobility, which are affected by the presence of impurities.

Specialized Vocabulary:

- direct current 直流、直流电
- exciton 激子
- Heterojunction 异质结
- conjugated system 共轭体系
- highest occupied molecular orbital (HOMO) 最高占有分子轨道
- lowest unoccupied molecular orbital (LUMO) 最低空分子轨道
- covalent bond 共价键
- Hydrocarbons 碳氢化合物，烃类
- delocalized bonding 离域键、非定域键
- antibonding orbital 反键轨道
- planar donor - acceptor heterojunction 平面供体-受体异质结
- diffusion length 扩散长度

6.2 Dye - sensitized solar cells (DSSCs)

The modern version of a dye solar cell, also known as the Grätzel cell, was originally co - invented in 1988 by Brian O'Regan and Michael Grätzel at UC Berkeley. Since 1991, when Grätzel and coworkers reported the development of the highly efficient, novel DSSCs, researchers throughout the world have intensively investigated DSSCs mechanisms, new materials, and commercialization. A maximum efficiency of 12.3% has been obtained under AM1.5 in the laboratory. In addition, satisfactory long - term stability of sealed cells has been achieved under relatively mild test conditions (low temperatures and no UV exposure). It is possible to achieve commercial DSSC production for indoor applications, such as calculators and several kinds of watches. Moreover, recently, a new DSSCs design achieved the indoor illumination power conversion efficiency as high as 28.9% under 1000 lux.

6.2.1 What is DSSC?

The DSSC is an attractive and promising device for solar cell applications that have been intensively investigated worldwide, and its PV mechanism is well understood. It is based on a semiconductor formed between a photo - sensitized anode and an electrolyte, a photoelectrochemical system. Commercial applications of the DSSC have been under intensive investigation for some years. The cost of commercially fabricating DSSC is expected to be relatively low because the cells are made of low - cost materials and assembly is simple and does not require high vacuum or high temperature.

When light falls onto the DSSC it is absorbed by the dye. The electrons that are excited, due to the extra energy the light provides, can escape from the dye and into the TiO_2 and diffuse through the TiO_2 to the electrode. They are eventually returned to the dye through the electrolyte. Therefore, DSSC uses a dye molecule to absorb light, similar with chlorophyll in photosynthesis. But the chlorophyll's not involved in charge transport. It just absorbs light and generates a charge, and then those charges are conducted by some well - established mechanisms. That's exactly what DSSC system does.

6.2.2 The structure of DSSC

The structure of DSSC mainly includes three parts, porous anode film on the conductive substrate which adsorbed dyes, electrolyte and counter electrode as a sandwich structure (Figure 6 - 3). The most important part is the porous photo - anode film where the electronic conduction takes place.

6.2.3 The primary processes and mechanism of DSSC

The schematic working diagram of DSSC is shown in Figure 6 - 4. The following primary steps convert photons to current. First, the photosensitizers adsorbed on the TiO_2

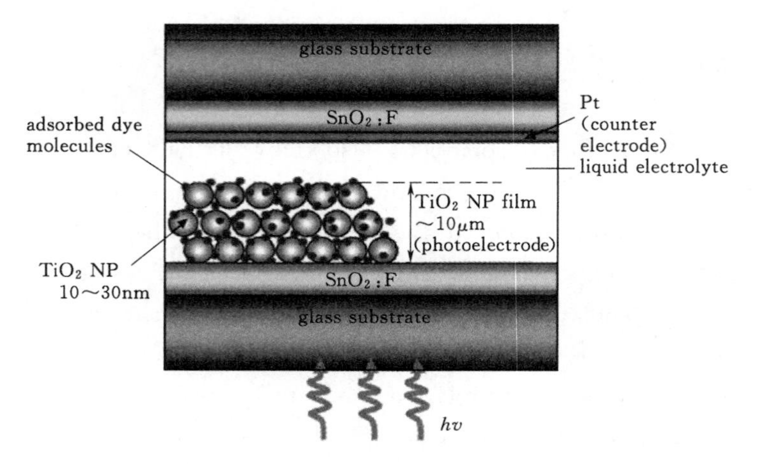

Figure 6 - 3 Structure of DSSC.

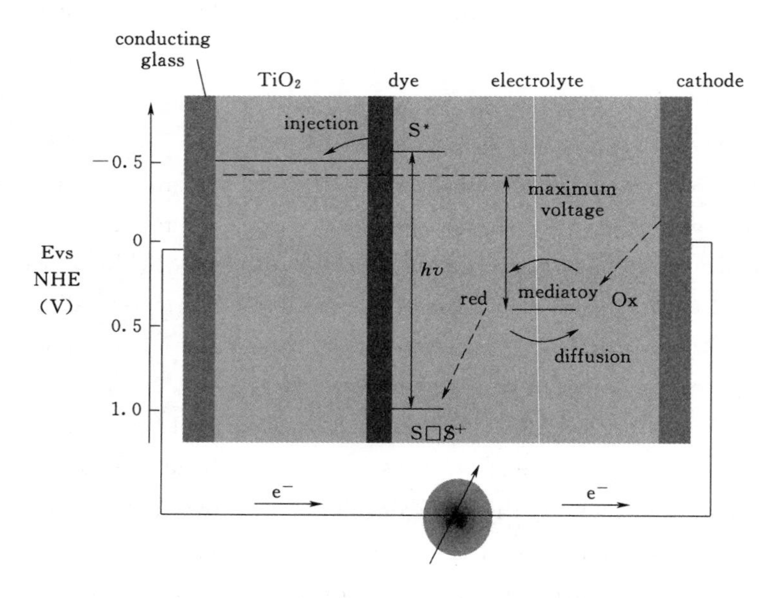

Figure 6 - 4 Schematic working diagram of DSSC.
(Source: Journal of photochemistry and photobiology C - photochemistry reviews, Michael Gratzel, 2003, 4, 145 - 153.)

surface absorb incident photons, then the photosensitizers are excited from the ground state (S) to the excited state (S^*) owing to the MLCT (metal to ligand charge transfer) transition. Second, the excited electrons are injected into the conduction band of the TiO_2 electrode, resulting in the oxidation of the photosensitizer (S^+).

$$S + hv \longrightarrow S^* \tag{6.1}$$

$$S^* \longrightarrow S^+ + e^-(TiO_2) \tag{6.2}$$

Injected electrons in the conduction band of TiO_2 are transported between TiO_2 nanoparticles with diffusion toward the back contact (TCO) and consequently reach the counter electrode through the external load and wiring. Third, the oxidized photosensitizer (S^+) accepts electrons from the I^- ion redox mediator, regenerating the ground state (S), and I^- is oxidized to the oxidized state, I^{3-}.

$$S^+ + e^- \longrightarrow S \quad (6.3)$$

Then, the oxidized redox mediator, I^{3-}, diffuses toward the counter electrode and is reduced to I^- ions.

$$I_3^- + 2e^- \longrightarrow 3I^- \quad (6.4)$$

Overall, electric power is generated without permanent chemical transformation.

In contrast to a conventional p-n type solar cell, the mechanism of a DSSC does not involve a charge-recombination process between electrons and holes because electrons are only injected from the photosensitizer into the semiconductor and a hole is not formed in the valence band of the semiconductor. In addition, charge transport takes place in the TiO_2 film, which is separated from the photon absorption site (i. e. the photosensitizer); thus, effective charge separation is expected. This photon-to-current conversion mechanism in a DSSC is similar to the mechanism for photosynthesis in nature, in which chlorophyll functions as the photosensitizer and charge transport occurs in the membrane.

6.2.4 Advantages

The biggest advantage of DSSC is that the charge transfer is accomplished by majority carrier transmission charge, there is no traditional p-n junction solar cell in minority carrier and charge transport materials surface recombination. In addition, its preparation process is simple, with simple requirements for the environmental conditions. Moreover, it has a potentially low cost (1/10~1/5 of silicon solar cells), long day life such as 15~20 years in lower light applications, and it can be mass produced because of the simple structure and easy of manufacture. In conclusion, the advantages are simple preparation and it is less demanding on the environment.

6.2.5 Problems

For expanded commercial applications, however, there are several problems for DSSC. Overcoming these problems would brings DSSC close to expanded commercialization. First of all, the efficiency of DSSC should be improved. For commercial applications, the full sun efficiency which is 12.3% now needs to be improved. Therefore, expanding the absorption property of the photosensitizer to near IR region is necessary to increase J_{SC} and resulting in efficiency improvement. Second, for outdoor applications, DSSC should also have long-term stability under more rigorous conditions, since it has already been tested only under relatively mild conditions. Third, development of solid-state electrolytes in DSSC is considered essential for developing a cell with long-term stability and is therefore critical for commercialization. However, it is very difficult to form a good solid-solid interface.

Specialized Vocabulary:

- anode/photoanode film 阳极/光阳极薄膜
- conductive substrate 导电基底
- counter electrode 对电极
- mesoporous 介孔、中孔
- anatase 锐钛矿（型）
- nanocrystalline 纳米晶（体）
- photosensitizers 光敏剂
- electrolyte 电解液，电解质
- redox 氧化还原，氧化还原剂，氧化还原反应
- charge recombination 电荷复合
- valence band 价带

6.3 Perovskite solar cells (PSCs)

In 2009, Miyasaka research group introduced perovskite materials ($CH_3NH_3PbI_3$ & $CH_3NH_3PbBr_3$) as light absorption film into the field of solar cells for the first time with photoelectric conversion of 3.8%. Since then, organometal halide perovskites have emerged as a promising light - harvesting material for high - efficiency solar cell devices. Perovskite sensitizer made a breakthrough in solid - state solar cells, where a rapid increase in efficiency has achieved at least 23.4%. As the most talked about, the most promising solar cells, its in - depth study also has important strategic significance.

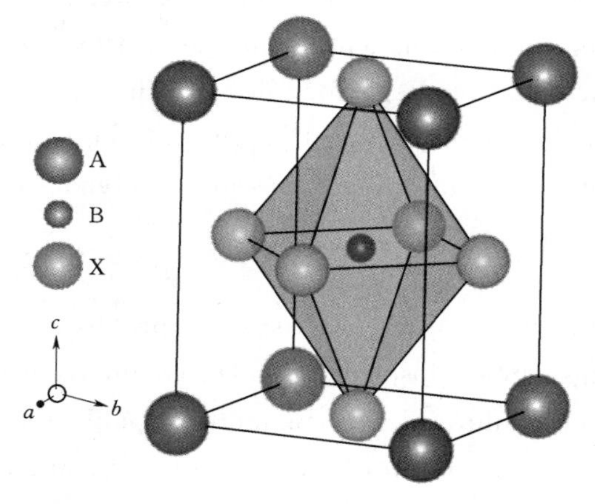

Figure 6 - 5 Crystal structure of the perovskite absorber adopting the perovskite ABX_3 form, where A is methylammonium, B is Pb and X is halogen.
(Source: MingZhen Liu, Michael B. Johnston, Henry J. Snaith, Efficient planar heterojunction perovskite solar cells by vapour deposition, Nature, 2013.)

An inorganic - organic lead halide perovskite is any material that crystallizes into an AMX_3 [where A is an organic ammonium cation ($CH_3NH_3^-$), M is Pb or Sn, and X is a halide anion such as Cl^-, Br^-, or I^-) structure (Figure 6 - 5, where A is methylammonium, B is Pb and X is halogen). The size of cation A is critical for the formation of a close - packed perovskite structure. In particular, cation A must fit into the space composed of four adjacent corner - sharing MX_6 octahedra. Of the various inorganic - organic lead halide perovskite materials, methylammonium lead iodide ($MAPbI_3$), with a bandgap of about 1.5 ～

1.6eV and a light absorption spectrum up to a wavelength of 800nm, has been extensively used as a light harvester in perovskite solar cells.

Generally, perovskite solar cells are composed of a transparent conductive substrate, a dense barrier layer, porous electron transport layer, perovskite absorption layer, hole transport layer and metal electrode, as shown in Figure 6 - 6. The material composition, microstructure and properties of the electron transport layer, the perovskite absorption layer and the hole transport layer have a significant effect on the photovoltaic performance and long term stability of the perovskite solar cells. At present, the perovskite solar cells have three main alternative structures. The first is mesoporous perovskite solar cells which commonly use mesoporous TiO_2 material as electron transport layer. By the impact of photons, there will be exciton separation phenomenon to form electrons and holes in the cell. Affected by the light - absorbing material, TiO_2 will transfer the electrons to the FTO surface. The second structure is planar perovskite solar cells. This type of perovskite solar cells is mainly developed for low temperature manufacture, with the advantages of low cost. When exposed to light, the cell will produce electron - hole pairs, thus make the electrons more active. When the circuit is connected, the electron - hole pairs will move rapidly, resulting in current, and then give full play to the role of perovskite solar cells. The third is flexible perovskite solar cells. Until 2015, the research of flexible perovskite solar cells in the world mainly used atomic layer vapor deposition manufacturing method, has boosted the efficiency of flexible perovskite solar cells to 18%.

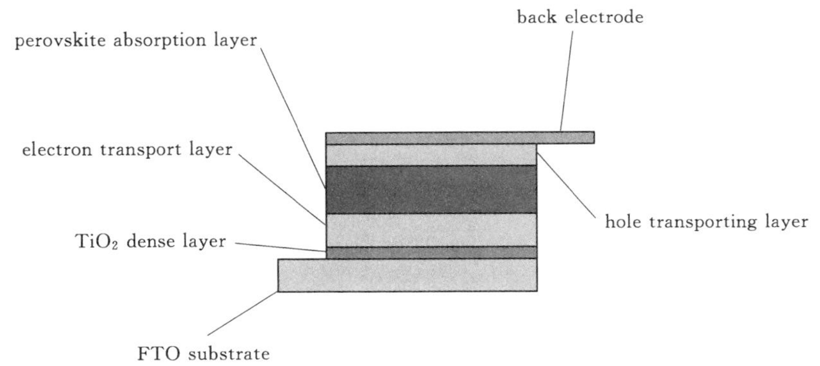

Figure 6 - 6 Schematic diagram of perovskite solar cell structure and the appearance of the cells.

The working principle of perovskite solar cell is shown in Figure 6 - 7. Methyl ammonium (MA) lead halide perovskite absorbs photons under light, and the electrons in the valence band jump to the conduction band, then the conduction band electrons are injected into the conduction band of TiO_2 and then transfer to the FTO conductive substrate. Meanwhile, holes are transferred to the organic hole transport layer, thus the electron - hole pairs are separated. When the external circuit is connected, the movement of electrons and holes will generate current. The main function of the dense layer is to collect the electrons which are injected into the electron transfer layer, resulting in the separation of the electron - hole pairs from the absorption layer. The main function of the absorption layer is to absorb the electron - hole pairs generated by sunlight and transfer the electron - hole pairs efficiently, thus electrons and holes reach respectively the corresponding dense layer and hole transport layer. The main role of hole transport layer is to collect and transfer holes which are injected from the absorption layer, and in the mean time to accelerate separation of electron - hole pairs in absorption layer together with dense layer.

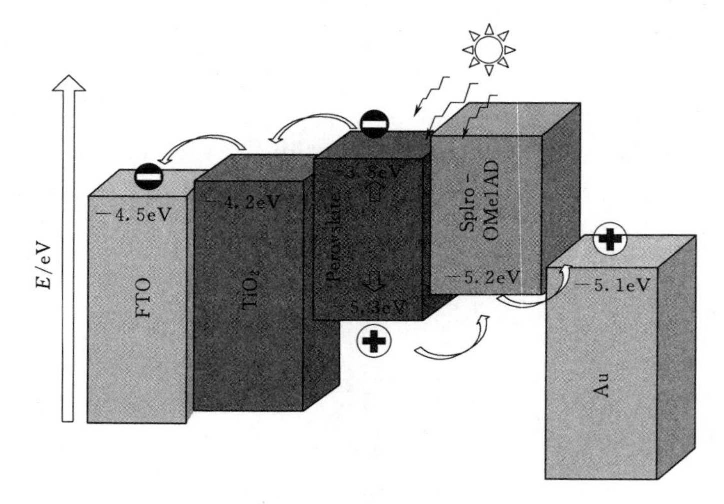

Figure 6 - 7 Working principle of perovskite solar cell.

On the preparation methods of perovskite solar cells, there are one - step spin - coating method, two - step sequential method, and vapor deposition in a high - vacuum chamber, plus other variants. Additionally, the fabrication methods for different device architectures such as planar and mesostructured cells are a little bit different.

At present, the development of perovskite solar cells is very vibrant, but there are still a number of key factors that may restrict the prospects of perovskite solar cells. First of all is the stability of the solar cell. The perovskite solar cells are sensitive and unstable in the air environment with sensitivity to oxygen and moisture, and the attenuation of efficiency for the MA compositions is rapid and significant. Second, the perovskite absorption layer contains soluble heavy metal lead, it is easy to cause environment pollution, though the concentration is low. Third, the most widely used fabrication method of perovskite materials is spin-coating, which is difficult to deposit in large areas and as a continuous film, so other methods need to be improved in order to obtain large area high efficiency perovskite solar cells to enable future commercial production. Finally, the theoretical study of perovskite solar cells has yet to be strengthened.

Specialized Vocabulary:

- organometal halide perovskites 有机金属卤化物钙钛矿
- inorganic-organic lead halide perovskite 有机-无机卤化铅钙钛矿
- dense layer 致密层
- electron transport layer 电子传输层
- perovskite absorption layer 钙钛矿吸收层
- hole transport layer 空穴传输层
- metal electrode 金属电极
- back electrode 背电极
- mesoporous perovskite solar cells 介孔钙钛矿太阳电池
- planar perovskite solar cells 平面/板钙钛矿太阳电池
- flexible perovskite solar cells 柔性钙钛矿太阳电池
- spin-coating 旋涂，旋涂仪
- atomic layer vapor deposition 原子层气相沉积

6.4 Quantum dot solar cells (QDSCs)

At present, there are quantum dot sensitized solar cells, quantum dot polymer hybrid solar cells, quantum dot schottky and depletion heterojunction quantum dot solar cells, etc. The highest efficiency of quantum dot sensitized solar cells is currently up to 13.4%.

The structure and working principle of quantum dot sensitized solar cells are similar to dye-sensitized solar cells. As shown in Figure 6-8, it mainly consists of a conductive substrate material (transparent conductive glass), wide band gap oxide semiconductor nanoporous film, quantum dots, electrolyte and counter electrode. Under the sun light, the quantum dots absorb the photons with the energy greater than the band gap width, which causes them to jump from the ground state to the excited state. The electrons in the excited state will be quickly injected into the conduction band of the nano wide band gap semiconductor, and then diffuse towards and transport to the conductive substrate, arriving to the counter electrode through external circuit. The quantum dots in the oxidized state are reduced and regenerated by the reducing agent in the electrolyte, and the oxidant in the electrolyte is reduced by receiving electrons at the counter electrode. The above is a cycle of electrons during the work mode of solar cells. However, in addition to this ideal electronic transmission, there is an inverse electron recombination process when the solar cells are actually working which is a limit to improvement of the cell performance.

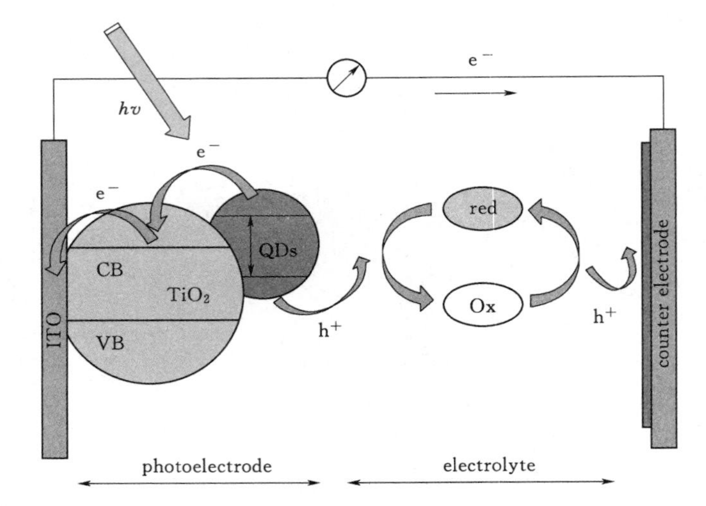

Figure 6-8 Structure and working principle of quantum dots solar cells.

Specialized Vocabulary:

- quantum dot sensitized solar cells 量子点敏化太阳电池
- quantum dot polymer hybrid solar cells 量子点聚合物杂化太阳电池
- quantum dot schottky solar cells 量子点肖特基太阳电池
- depletion heterojunction quantum dot solar cells 耗尽异质结量子点太阳电池
- Inverse electron recombination 电子反向复合过程

6.5 Reading material (Please read the article and find out the specialized vocabulary)

6.5.1 Efficient planar heterojunction perovskite solar cells by vapour deposition (From Nature, 501, 396-399, 2013)

Many different photovoltaic technologies are being developed for large-scale solar energy conversion. The wafer-based first generation photovoltaic devices have been followed by thin-film solid semiconductor absorber layers sandwiched between two charge-selective contacts and nanostructured (or mesostructured) solar cells that rely on a distributed heterojunction to generate charge and to transport positive and negative charges in spatially separated phases. Although many materials have been used in nanostructured devices, the goal of attaining high-efficiency thin film solar cells in such a way has yet to be achieved. Organometal halide perovskites have recently emerged as a promising material for high-efficiency nanostructured devices. Here we show that nanostructuring is not necessary to achieve high efficiencies with this material: a simple planar heterojunction solar cell incorporating vapour - deposited perovskite as the absorbing layer can have solar - to - electrical power conversion efficiencies of over 15 percent (as measured under simulated full sunlight). This demonstrates that perovskite absorbers can function at the highest efficiencies in simplified device architectures, without the need for complex nanostructures.

Within a solar cell there are many different components with discrete roles and having different tolerances for purity and optoelectronic properties. The hybrid inorganic - organic solar cell concept is "material agnostic" in that it aims to use the optimum material for each individual function. Any material that is easy to process, inexpensive and abundant can be used, with the aim of delivering a high - efficiency solar cell. Hybrid solar cells have been demonstrated in p - conjugated polymer blends containing semiconductor nanocrystals such as CdSe, $CuInS_2$ and PbS. Dye - sensitized solar cells are hybrid solar cells containing a mesostructured inorganic n - type oxide (such as TiO_2) sensitized with an organic or metal complex dye, and infiltrated with an organic p - type hole - conductor. Recently, organometal trihalide perovskite absorbers with the general formula $(RNH_3)BX_3$ (where R is C_nH_{2n+1}, X is the halogen I, Br or Cl, and B is Pb or Sn), have been used instead of the dye in dye - sensitized solar cells to deliver solid - state solar cells with a power conversion efficiency of over 10%.

Evolving from the dye - sensitized solar cells, we found that replacing the mesoporous TiO_2 with mesoporous Al_2O_3 resulted in a significant improvement in efficiency, delivering an open - circuit voltage of over 1.1V in a device which we term a "meso - superstructured solar cell". We reason that this observed enhancement in open - circuit voltage is due to confinement of the photo - excited electrons within the perovskite phase, thereby increasing the splitting of the quasi - Fermi levels for electrons and holes under illumination, which is ultimately responsible for generating the open - circuit voltage. Further removal of the thermal sintering of the mesoporous Al_2O_3 layer, and better optimization of processing, has led to meso - superstructured solar cells with more than 12% efficiency. In addition, $CH_3NH_3PbI_{3-x}Cl_x$ can operate relatively efficiently as a thin - film absorber in a solution - processed planar heterojunction solar cell configuration, delivering around 5% efficiency when no mesostructure is involved. This previous work demonstrates that the perovskite absorber is capable of operating in a much simpler planar architecture, but raises the question of whether mesostructure is essential for the highest efficiencies, or whether the thin - film planar heterojunction will lead to a superior technology.

Here, as a means of creating uniform flat films of the mixed halide perovskite $CH_3NH_3PbI_{3-x}Cl_x$, we use dual - source vapour deposition. In Figure 6 - 9 we show an illustration of the vapour - deposition set - up, along with an illustration of a planar heterojunction p - i - n solar cell [Figure 6 - 9 (c)]. From the bottom (the side from which the light is incident), the device is constructed on fluorine - doped tin oxide (FTO) - coated glass, coated with a compact layer of n - type TiO_2 that acts as the electron selective contact. The perovskite layer is then deposited on the n - type compact layer, followed by the p - type hole conductor, 2·, 29, 7, 79 - tetrakis - (N, N - di - p - methoxyphenylamine) 9, 99 - spiro-bifluorene (spiro - OMeTAD), which ensures the selective collection of holes at the silver cathode. Given that the purpose of this study was to understand and optimize the properties of the vapour - deposited perovskite absorber layer, the compact TiO_2 and the spiro - OMeTAD hole transporter were solution processed, as is usual in meso - superstructured solar cells.

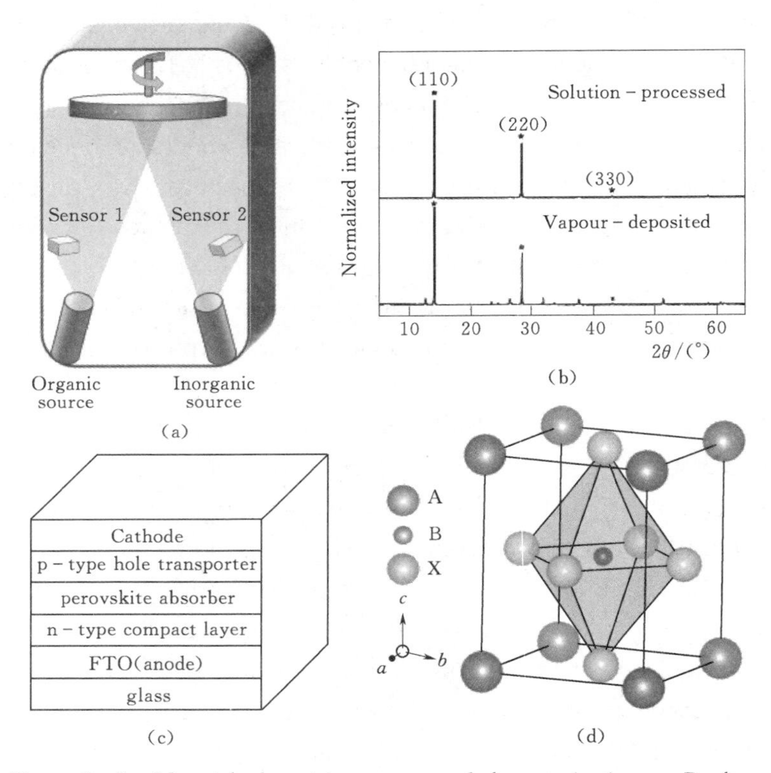

Figure 6 - 9 Materials deposition system and characterization. a, Dual - source thermal evaporation system for deposition the perovskite absorbers; the organic source was methylammonium iodide and the inorganic source $PbCl_2$. b, X - ray diffraction spectra of a solution - processed perovskite film (blue) and vapour - deposited perovskite film (red). The baseline is offset for ease of comparision and the intensity has been normalized. c, Generic structure of a planar heterojunction p - i - n perovskite solar cell. d, Crystal structure of the pervoskite absorber adopting the perovskite ABX_3 form, where A is methylammonium, B is Pb and X is I or Cl.

In Figure 6 - 9 (b), we compare the X - ray diffraction pattern of films of $CH_3NH_3PbI_{3-x}Cl_x$ either vapour - deposited or solution - cast onto compact TiO_2 - coated FTO - coated glass. The main diffraction peaks, assigned to the 110, 220 and 330 peaks at 14.12°, 28.44° and, respectively, 43.23°, are in identical positions for both solution - processed and vapour - deposited films, indicating that both techniques have produced the same mixed - halide perovskite with an orthorhombic crystal structure. Notably, looking closely in the region of the (110) diffraction peak at 14.12°, there is only a small signature of a peak at 12.65° [the (001) diffraction peak for PbI_2] and no measurable peak at 15.68° [the (110) diffraction peak for $CH_3NH_3PbCl_3$], indicating a high level of phase purity. A diagram of the crystal structure is shown in Figure 6 - 10 (d). The main difference between $CH_3NH_3PbI_3$ and the mixed - halide perovskite presented here is evident in a slight contraction of the c axis. This is consistent with the Cl atoms in the mixed - halide perovskite residing in the apical positions, out of the PbI_4 plane, as opposed to in the equatorial octahedral sites, as has been theoretically predicted.

We now make a comparison between the thin - film topology and cross sectional structure of devices fabricated by either vapour deposition or solution processing. The top - view scanning electron microscope (SEM) images in Figure 6 - 10 (a), (b) highlight the consid-

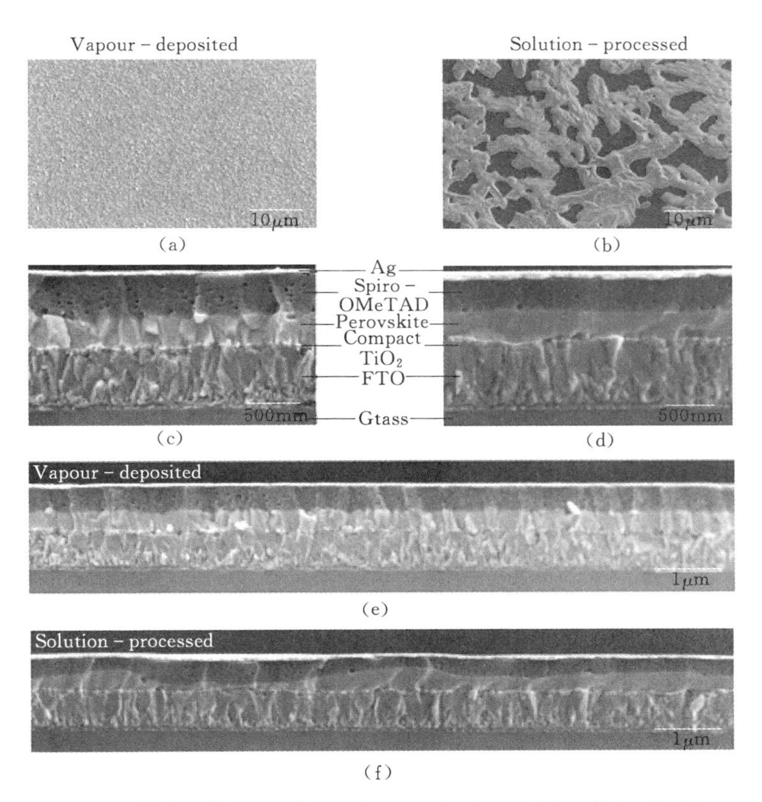

Figure 6 - 10 Thin - film topology characterization. (a), (b), SEM top views of a vapour - deposited perovskite film (a) and a solution - processed perovskite film (b). (c), (d), Cross - sectional SEM images under high magnification of complete solar cells constructed from a vapour - deposited perovskite film (c) and a solution - processed perovskite film (d). (e), (f), Cross - sectional SEM images under lower magnification of completed solar cells constructed from a vapour - deposited perovskite film (e) and a solution - processed perovskite film (f).

erable differences between the film morphologies produced by the two deposition processes. The vapour - deposited films are extremely uniform, with what appear to be crystalline features on the length scale of hundreds of nanometers. In contrast, the solution - processed films appear to coat the substrate only partially, with crystalline "platelets" on the length scale of tens of micrometers. The voids between the crystals in the solution - processed films appear to extend directly to the compact TiO_2 - coated FTO coated glass.

The cross - sectional images of the completed devices in Figure 6 - 10 (c), d reveal more information about the crystal size. The vapour - deposited perovskite film [Figure 6 - 10 (c)] is uniform and similar in appearance to the FTO layer, albeit with slightly larger crystal features. The solution processed perovskite film [Figure 6 - 10 (d)] is extremely smooth in the SEM image, consistent with much larger crystal grain size than the field of view. For both of these films the crystal sizes are larger than we are able to determine from the peak width of the X - ray diffraction spectra (about 400nm), owing to machine broadening. On zooming out, the vapour - deposited film in Figure 6 - 10 (e) remains flat, having an average film thickness of approximately 330nm. In contrast, the solution - processed film in Figure 6 - 10 (f) has an undulating nature, with film thickness varying from 50 to 410nm. Notably, this cross - section is still within a single 'platelet', and so even greater long - range roughness occurs owing to the areas where the perovskite absorber is completely absent (a thickness variation from 0 to 465nm was observed in multiple SEM images).

The current - density/voltage curves measured under simulated AM1. 5101mW • cm^{-2} irradiance (simulated sunlight) for the best - performing vapour - deposited and solution - processed planar heterojunction solar cells are shown in Figure 6 - 11. The most efficient vapour - deposited perovskite device had a short - circuit photocurrent of 21. 5mA • cm^{-2}, an open - circuit voltage of 1. 07V and a fill factor of 0. 68, yielding an efficiency of 15. 4%. In the same batch, the best solution - processed planar heterojunction perovskite solar cell produced a short - circuit photocurrent of 17. 6mA • cm^{-2}, an open - circuit voltage of 0. 84V and a fill factor of 0. 58, yielding an overall efficiency of 8. 6%. In Table 6 - 1 we show the extracted performance parameters for these best - performing cells and the average with standard deviation of a batch of 12 vapour deposited perovskite solar cells fabricated in an identical manner to the best - performing cell.

Table 6 - 1 Solar cell performance parameters.

	Current density /(mA • cm^{-2})	Open - circuit voltage/V	Fill factor	Efficiency /%
Vapour - deposited	21. 5	1. 07	0. 67	15. 4
Vapour - deposited (average +- s. d.)	18. 9+-1. 8	1. 05+-0. 03	0. 62+-0. 05	12. 3+-2. 0
Solution - processed	17. 6	0. 84	0. 58	8. 6

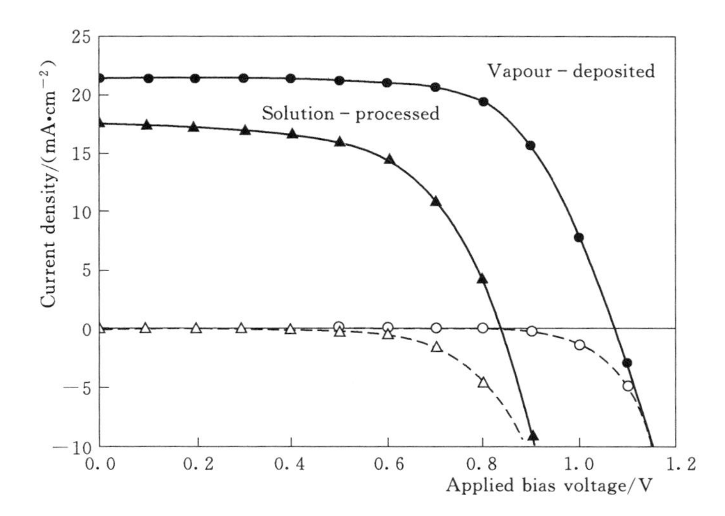

Figure 6 - 11 Solar cell performance. Current - density/voltage curves of the best - performing solution - processed (blue lines, triangles) and vapour - deposited (red lines, circles) planar heterojunction perovskite solar cells measure under simulated AM1.5 sunlight of $101mW \cdot cm^{-2}$ irradiance (solid lines) and in the dark (dashed lines). The curves are for the best - performing cells measured and their reproducibility is shown in Table 6 - 1.

Dual - source vapour deposition results in superior uniformity of the coated perovskite films over a range of length scales, which subsequently results in substantially improved solar cell performance. In optimizing planar heterojunction perovskite solar cells, perovskite film thickness is a key parameter. If the film is too thin, then that region will not absorb sufficient sunlight. If the film is too thick, there is a significant chance that the electron and hole (or exciton) diffusion length will be shorter than the film thickness, and that the charge will therefore not be collected at the p - type and n - type heterojunctions. Furthermore, the complete absence of material from some regions in the solution - processed films (pinholes) will result in direct contact of the p - type spiro - OMeTAD and the TiO_2 compact layer. This leads to a shunting path that is probably partially responsible for the lower fill factor and open - circuit voltage in the solution - cast planar heterojunction devices. Indeed, it is remarkable that such inhomogeneous and undulating solution - cast films can deliver devices with over 8% efficiency.

The results presented here demonstrate that solid perovskite layers can operate extremely well in a solar cell, and in essence set a lower limit of 330nm (the film thickness) on the electron and hole diffusion length in this perovskite absorber. However, more work is required to determine the electron and hole diffusion lengths precisely and to understand the primary excitation and the mechanisms for free - charge generation in these materials.

A distinct advantage of vapour deposition over solution processing is the ability to prepare layered multi-stack thin films over large areas. Vapour deposition is a mature technique used in the glazing industry, the liquid-crystal display industry and the thin-film solar cell industry, among others. Vapour deposition can lead to full optimization of electronic contact at interfaces through multilayers with controlled levels of doping, as is done in the crystalline silicon "heterojunction with thin intrinsic layer" solar cell and in thin-film solar cells. Additionally, organic light-emitting diodes have proved to be commercially sound, with devices with extremely thin multilayer stacks fabricated by vapour deposition. Small molecular organic photovoltaics have also been able to compete directly with solution-processed organic photovoltaics despite much lower levels of research and development, because with vapour deposition the charge-collection interfaces can be carefully tuned, and multi-junction architectures are more straightforward to realize. An interesting possibility for the current vapour-deposited perovskite technology is to use it as a "top cell" in a hybrid tandem junction with either crystalline silicon or copper indium gallium (di) selenide. Although ultimately an "all-perovskite" multi-junction cell should be realizable, the perovskite cells have now achieved a performance that is sufficient to increase the absolute efficiency of high-efficiency crystalline silicon and copper indium gallium (di) selenide solar cells. Additionally, because vapour deposition of the perovskite layers is entirely compatible with conventional processing methods for silicon wafer-based and thin-film solar cells, the infrastructure could already be in place to scale up this technology.

We have built vapour-deposited organometal trihalide perovskite solar cells based on a planar heterojunction thin-film architecture that have a solar-to-electrical power conversion efficiency of over 15% with an open-circuit voltage of 1.07V. The perovskite absorbers seem to be versatile materials for incorporation into highly efficient solar cells, given the low-temperature processing they require, the option of using either solution processing or vapour deposition or both, the simplified device architecture and the availability of many other metal and organic salts that could form a perovskite structure. Whether vapour deposition emerges as the preferred route for manufacture or simply represents a benchmark method for fabricating extremely uniform films (that will ultimately be matched by solution processing) remains to be seen. Finally, a key target for the photovoltaics community has been to find a wider-bandgap highly efficient "top cell", to enable the next step in improving the performance of crystalline silicon and existing second-generation thin-film solar cells. This perovskite technology is now compatible with these first- and second-generation technologies, and is hence likely to be adopted by the conventional photovoltaics community and industry. Therefore, it may find its way rapidly into utility-scale power generation.

6.5.2 Dye - sensitized solar cells

(From Journal of photochemistry and photobiology C - photochemistry reviews, Michael Gratzel 2003, 4, 145 - 153)

1. *Introduction*

Photovoltaic devices are based on the concept of charge separation at an interface of two materials of different conduction mechanism. To date this field has been dominated by solid - state junction devices, usually made of silicon, and profiting from the experience and material availability resulting from the semiconductor industry. The dominance of the photovoltaic field by inorganic solid - state junction devices is now being challenged by the emergence of a third generation of cells, based, for example, on nanocrystalline and conducting polymers films. These offer the prospective of very low cost fabrication and present attractive features that facilitate market entry. It is now possible to depart completely from the classical solid - state junction device, by replacing the contacting phase to the semiconductor by an electrolyte, liquid, gel or solid, thereby forming a photo - electrochemical cell. The phenomenal progress realized recently in the fabrication and characterization of nanocrystalline materials has opened up vast new opportunities for these systems. Contrary to expectation, devices based on interpenetrating networks of mesoscopic semiconductors have shown strikingly high conversion efficiencies, which compete with those of conventional devices. The prototype of this family of devices is the dye - sensitized solar cell, which realizes the optical absorption and the charge separation processes by the association of a sensitizer as light - absorbing material with a wide band gap semiconductor of nanocrystalline morphology.

2. *Operation principle of the dye - sensitized nanocrystalline solar cell (DSC)*

A schematic presentation of the operating principles of the DSC is given in Figure 6 - 12. At the heart of the system is a mesoporous oxide layer composed of nanometer - sized particles which have been sintered together to allow for electronic conduction to take place. The material of choice has been TiO_2 (anatase) although alternative wide band gap oxides such as ZnO, and Nb_2O_5 have also been investigated. Attached to the surface of the nanocrystalline film is a monolayer of the charge transfer dye. Photo excitation of the latter results in the injection of an electron into the conduction band of the oxide. The original state of the dye is subsequently restored by electron donation from the electrolyte, usually an organic solvent containing redox system, such as the iodide/triiodide couple. The regeneration of the sensitizer by iodide intercepts the recapture of the conduction band electron by the oxidized dye. The iodide is regenerated in turn by the reduction of triiodide at the counter electrode the circuit being completed via electron migration through the external load. The voltage generated under illumination corresponds to the difference between the Fermi level of the electron in the solid and the redox potential of the electrolyte. Overall the device generates electric power from light without suffering any permanent chemical transformation.

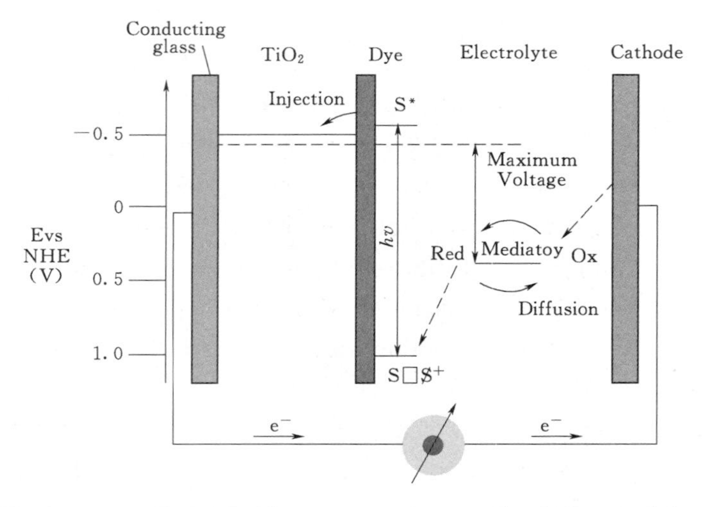

Figure 6 - 12　Principle of operation and energy level scheme of the dye - sensitized nanocrystalline solar cell. Photo - excitation of the sensitizer (S) is followed by electron injection into the conduction band of the mesoporous oxide semiconductor. The dye molecule is regenerated by the redox system, which itself is regenerated at the counter electrode by electrons passed through the load. Potentials are referred to the normal hydrogen electrode (NHE). The open - circuit voltage of the solar cell corresponds to the difference between the redox potential of the mediator and the Fermilevel of the nanocrystallline film indicated with a dashed line.

Figure 6 - 13 shows the scanning electron micrograph of a typical TiO_2 (anatase) film deposited by screen printing on a conducting glass sheet that serves as current collector. The film thickness is typically $5 \sim 20$m and the TiO_2 mass about $1 \sim 4mg/cm^2$. Analysis of the layer morphology shows the porosity to be about $50\% \sim 65\%$, the average pore size being 15nm. The prevailing structures of the anatase nanoparticles are square - bipyramidal, pseudo cubic and stab like. According to HRTEM measurements the (101) face is mostly exposed followed by (100) and (001) surface orientations.

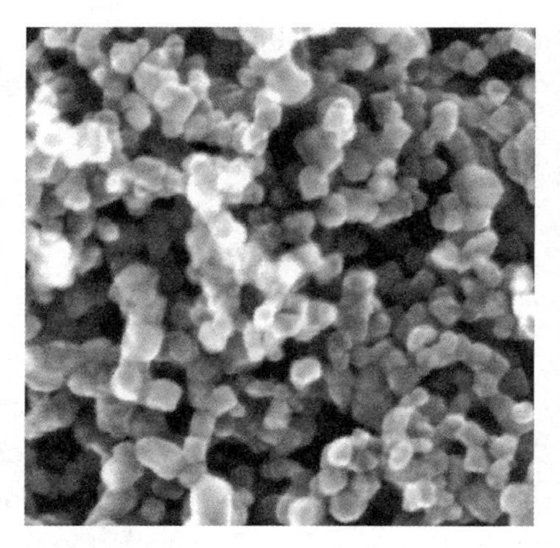

Figure 6 - 13　Scanning electron microscope picture of a nanocrystalline TiO_2 (anatase) film used in DSSC.

A recent alternative embodiment of the DSC concept is the sensitized heterojunction usually with an inorganic wide band gap nanocrystalline semiconductor of n-type polarity as electron acceptor, the charge neutrality on the dye being restored by a hole delivered by the complementary semiconductor, inorganic or organic and of p-type polarity. The prior photo-electrochemical variant, being further advanced in development, has an AM1.5 solar conversion efficiency of over 10%, while that of the solid-state device is, as yet, significantly lower.

3. *Historical background*

The history of the sensitization of semiconductors to light of wavelength longer than that corresponding to the bandgap has been presented elsewhere. It is an interesting convergence of photography and photo-electrochemistry, both of which rely on photo-induced charge separation at a liquid-solid interface. The silver halides used in photography have band gaps of the order of 2.7~3.2eV, and are therefore insensitive to much of the visible spectrum, just as is the TiO_2 now used in these photo-electrochemical devices.

The first panchromatic film, able to render the image of a scene realistically into black and white, followed on the work of Vogel in Berlin after 1873, in which he associated dyes with the halide semiconductor grains. The first sensitization of a photo-electrode followed shortly thereafter, using a similar chemistry. However, the clear recognition of the parallelism between the two procedures, a realization that the same dyes in principle can function in both and a verification that their operating mechanism is by injection of electrons from photo-excited dye molecules into the conduction band of the n-type semiconductor substrates date to the 1960s. In subsequent years the idea developed that the dye could function most efficiently if chemisorbed on the surface of the semiconductor. The concept emerged to use dispersed particles to provide a sufficient interface, then photo-electrodes where employed.

Titanium dioxide became the semiconductor of choice. The material has many advantages for sensitized photo-chemistry and photo-electrochemistry: it is a low cost, widely available, non-toxic and biocompatible material, and as such is even used in health care products as well as domestic applications such as paint pigmentation. The standard dye at the time was tris (2, 2^L-bipyridyl-4, 4^L-carboxylate) ruthenium (Ⅱ), the function of the carboxylate being the attachment by chemisorption of the chromophore to the oxide substrate. Progress thereafter, until the announcement in 1991 of the sensitized electrochemical photovoltaic device with a conversion efficiency at that time of 7.1% under solar illumination, was incremental, a synergy of structure, substrate roughness and morphology, dye photophysics and electrolyte redox chemistry. That evolution has continued progressively since then, with certified efficiency now over 10%.

Exercises and Discussion

1. Please describe the different types of organic solar cells with different structures.
2. Please describe the working mechanism and structure of DSSC.
3. What's the big difference between DSSC and silicon based solar cells?
4. Please describe the advantages of the perovskite solar cells.

Chapter 7

Solar Cell Modules

7.1 The definition and types of solar cell modules

Solar cell module is a minimum combination of integral solar cell device that have external packaging and internal connection, and can provide individual DC output. Solar cell modules are formed by multiple single solar cell interconnected encapsulation. Solar cell module is the core part of solar power generation systems and the most important part of the solar system.

A PV module consists of a number of interconnected solar cells encapsulated into a single, long - lasting, stable unit. The key purpose of encapsulating a set of electrically connected solar cells is to protect them and their interconnecting wires from the typically harsh environment in which they are used. For example, solar cells, since they are relatively thin, are prone to mechanical damage unless protected. In addition, the metal grid on the top surface of the solar cell and the wires interconnecting the individual solar cells may be corroded by water or water vapor. The two key functions of encapsulation are to prevent mechanical damage to the solar cells and to prevent water or water vapor from corroding the electrical contacts. Moreover, in order to meet the requirements of the load, series and/or parallel connection is required to form a minimum unit to be used as power supply independently. The single solar cell needs to supply electric power when in series or parallel connection because the voltage output of the single solar cell is hard to meet the normal demand.

There are several types of solar cell modules. According to the types of the solar cells, there are crystalline silicon (monocrystalline or multicrystalline silicon) solar cell modules, amorphous silicon thin film solar cell modules and gallium arsenide solar cell modules, etc (Figure 7 - 1) . According to the encapsulation materials and process, there are epoxy resin encapsulation panels and laminated packaging solar cell modules, etc. According to the application, there are standard solar cell modules and building materials type solar cell modules. Among them, building materials type solar cell modules are divided into single glass with back face opaque to transparent, double sided laminated glass module components and double hollow glass module components. The modules based on crystalline silicon solar cells accounted for more than 85% of the market.

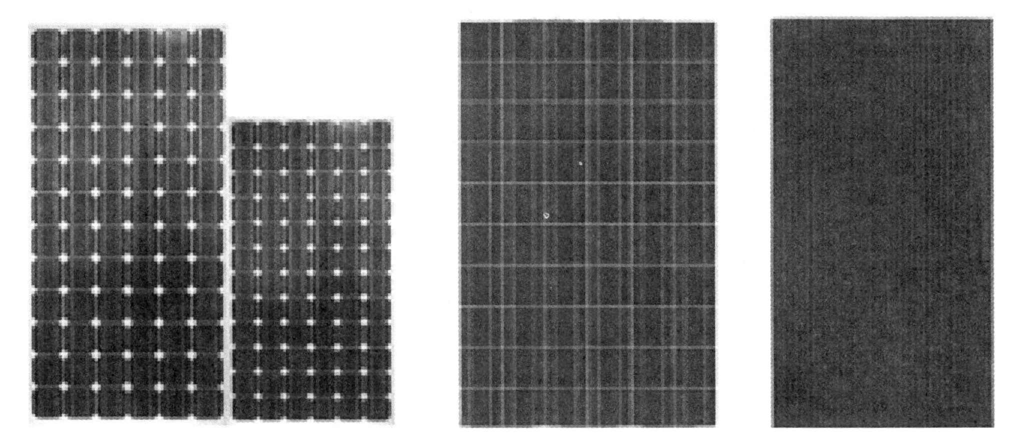

Figure 7 - 1 Images of monocrystalline silicon modules, polycrystalline silicon modules and amorphous silicon modules.

Specialized Vocabulary:

- modules 组件，模组，模块
- epoxy resin encapsulation 环氧树脂封装
- laminated packaging 层压封装

7.2 Module structure

Most bulk silicon PV modules consist of a transparent top surface, an encapsulant, a rear layer and a frame around the outer edge. In most modules, the top surface is glass, the encapsulant is ethyl vinyl acetate (EVA) and the rear layer is Tedlar, as shown in Figure 7 - 2.

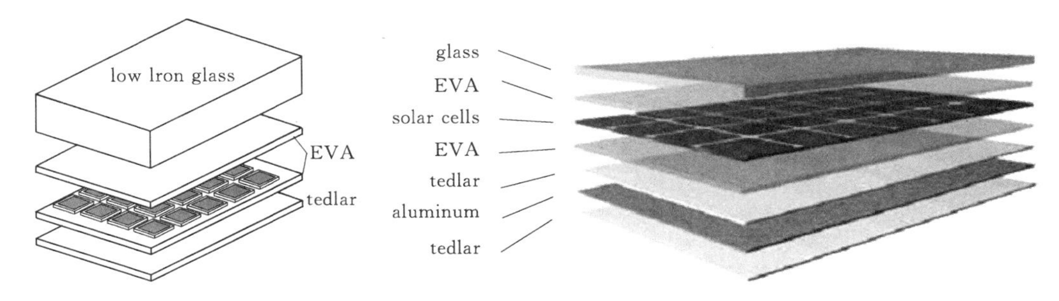

Figure 7 - 2 Typical bulk silicon module materials.

The front surface of a PV module must have a high transmission in the wavelengths which can be used by solar cells in the PV module. For silicon solar cells, the top surface must have high transmission of light in the wavelength range of 380nm to 1200nm. In addition, the reflection from the front surface should be low. There are several choices for a top surface material including acrylic, polymers and glass. Tempered, low iron - content glass is the most commonly used as it is low cost, strong, stable, highly transparent in the visible range, impervious to water and gases and has good self - cleaning properties. An encapsulant is used to provide adhesion between the solar cells, the top surface and the rear surface of the PV module. The encapsulant should be stable at elevated temperatures and high UV exposure. It should also be optically transparent and should have a low thermal resistance. EVA is the most commonly used encapsulant material. EVA comes in thin sheets which are inserted between the solar cells and the top surface and the rear surface. This sandwich is then heated to 150℃ to polymerize the EVA and bond the module together. The key characteristics of the rear surface of the PV module are that it must have low thermal resistance and that it must prevent the ingress of water or water vapor. In most modules, a thin polymer sheet, typically Tedlar, is used as the rear surface. Some PV modules, known as bifacial modules are designed to accept light from either the front or the rear of the solar cell. In bifacial modules both the front and the rear must be optically transparent. A final structural component of the module is the edging or framing of the module. A conventional PV module frame is typically made of aluminium. The frame structure should be free of projections which could result in the inclusion of water, dust or other matter. Photovoltaic module consists of transparent front side, encapsulated solar cells and backside. While front side material (superstrate) usually low - iron, tempered glass, for some special module types some other front side materials used include polycarbonate or non - tempered glass. For flexible modules ethylene tetrafluoroethylene (ETFE) a fluorine based plastic, with high corrosion resistance and strength over a wide temperature range is often used. The backside is usually non transparent, the most common used material is polyvinyl fluoride (PVF). Transparent back side is also possible - transparent back side materials are often used in modules that are integrated into buildings envelope (facade or roof), see also BIPV section. Between the backside glass and the solar cells the encapsulation materials are placed. Many different materials can be used for encapsulation but two materials most often used are EVA and polyvinyl - butiral (PVB). PVB is also used as a laminate in safety windscreens in automotive industry. It is used as encapsulation material in transparent modules. EVA is used for encapsulation of cells in standard modules. Other less common encapsulation materials are thermoplastic polyurethane (TPU) and castable polyurethane or silicone cast resins used for examples in transparent modules or other demanding applications. Required mechanical characteristics (impact resistance etc.) and module qualification procedures are defined in international standards.

Standard PV modules are laminates composed of a 4mm thick glass superstrate, an 0.5mm thick encapsulating polymer layer (EVA), a 0.18mm thick layer of Si solar cells, another layer of EVA with the same thickness as the previous one, and finally a thin multi-layered back sheet made of Tedlar/Aluminum/Tedlar 0.1mm thick, as shown in Figure 7-2. The majority of solar cells available on the market are made of either mono or polycrystalline Si between thin layers of EVA in the plane of the cells. Two main conductors, called busbars, connect the cells together and are placed on the upper and lower sides of the cells.

7.3 Encapsulation technology

It is difficult to meet the demand of conventional electricity due to the very small output power of single solar cell. Therefore the solar cells need to be encapsulated into modules to improve the power output. Encapsulation is a critical step in solar cell manufacturing. The encapsulation of the cells is not only the process for the module but also the guarantee of cell lifetime, moreover enhance the resistance of the cell to degradation.

EVA encapsulation technology is the most popular. The encapsulation process is: first, use glass as the substrate, and then place EVA films on both sides of the solar cell. The encapsulation process involves heating to a certain temperature under vacuum, so that the EVA softens, and as the temperature decreases, solidification occurs, building the solar cells in a fixed pattern. The second stage involves coating the connector materials at the edge of the base plate and the top plate of the module, and adding the frame. The third phase involves release testing of the modules, cleaning, packing and putting in storage. The schematic diagram of encapsulation process is in below in Figure 7-3.

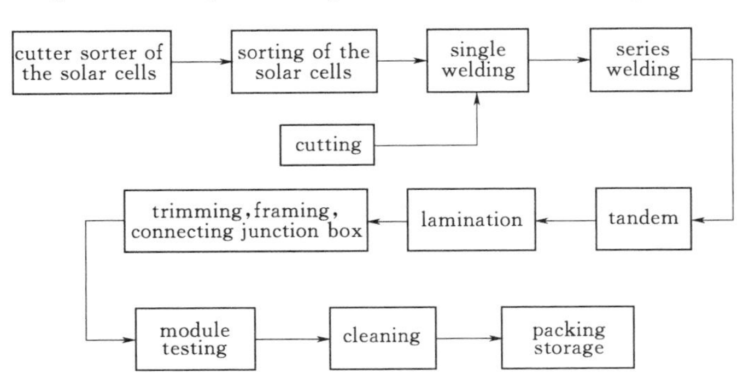

Figure 7-3 EVA copolymer encapsulation process.

Specialized Vocabulary:

- encapsulant 密封材料，密封剂
- a rear layer 底板
- ethyl vinyl acetate (EVA) 乙烯基醋酸乙脂
- Tedlar 泰德拉（宇宙飞船用以保留热量的绝缘材料），聚氟乙烯
- ethylene tetrafluoroethylene (ETFE) 聚氟乙烯

- polyvinyl fluoride (PVF) 聚氟乙烯
- encapsulation material 封装材料
- busbars 母线
- superstrate 覆盖物
- polyvinyl - butiral (PVB) 聚乙烯醇缩丁醛
- thermoplastic polyurethane (TPU) 热塑性聚氨酯
- power output 输出功率

7.4 Reading material (Please read the article and find out the specialized vocabulary)

Solar Electric Photovoltaic Modules (From Solar Direct)

Photovoltaic (PV) Power

PV is emerging as a major power resource, steadily becoming more affordable and proving to be more reliable than utilities. Photovoltaic power promises a brighter, cleaner future for our children.

Using the technology we have today we could equal the entire electric production of the United States with photovoltaic power plants using only about 12, 000 square miles.

In 1839, Edmund Becquerel discovered the process of using sunlight to produce an electric current in a solid material, but it wasn't until a century later that scientists eventually learned that the photovoltaic effect caused certain materials to convert light energy into electrical energy.

The photovoltaic effect is the basic principal process by which a PV cell converts sunlight into electricity. When light shines on a PV cell, it may be reflected, absorbed, or pass right through. The absorbed light generates electricity.

In the early 1950s, photovoltaic (PV) cells were developed as a spin - off of transistor technology. Very thin layers of pure silicon are impregnated with tiny amounts of other elements. When exposed to sunlight, small amounts of electricity are produced. Originally this technology was a costly source of power for satellites but it has steadily come down in price making it affordable to power homes and businesses.

Cells	Semiconductor device that converts sunlight into direct current (DC) electricity
Module	PV modules consist of PV cell circuits sealed in an environmentally protective laminate and are the fundamental building block of PV systems
Panels	PV panels include one or more PV modules assembled as a pre - wired, field - installable unit

From Cell to Array

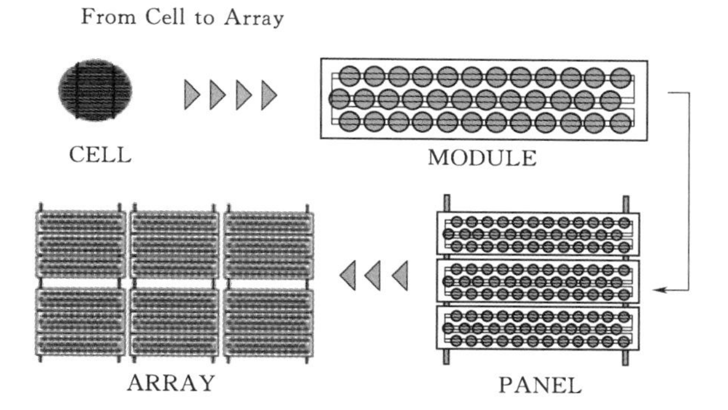

Array A PV array is the complete power-generating unit, consisting of any number of PV modules and panels

Photovoltaic cell

A single PV cell is a thin semiconductor wafer made of two layers generally made of highly purified silicon (PV cells can be made of many different semiconductors but crystalline silicon is the most widely used). The layers have been doped with boron on one side and phosphorous on the other side, producing surplus of electrons on one side and a deficit of electrons on the other side. When the wafer is bombarded by sunlight, photons in the sunlight knock off some of excess electrons, this makes a voltage difference between the two sides as the excess electrons try to move to the deficit side. In silicon this voltage is.5 volt. Metallic contacts are made to both sides of the semiconductor. With an external circuit attached to the contacts, the electrons can get back to where they came from and a current flows through the circuit. This PV cell has no storage capacity, it simply acts as an electron pump. The amount of current is determined by the number of electrons that the solar photons knock off. Bigger cells, more efficient cells, or cells exposed to more intense sunlight will deliver more electrons.

Photovoltaic Modules

A PV module consists of many PV cells wired in parallel to increase current and in series to produce a higher voltage. 36 cell modules are the industry standard for large power production. The module is encapsulated with tempered glass (or some other transparent material) on the front surface, and with a protective and waterproof material on the back surface. The edges are sealed for weatherproofing, and there is often an aluminum frame holding everything together in a mountable unit. In the back of the module there is a junction box, or wire leads, providing electrical connections. There are currently four commercial production technologies for PV Modules:

Single Crystalline

This is the oldest and more expensive production technique, but it's also the most efficient sunlight conversion technology available. Module efficiency averages about 10% to

12%.

Polycrystalline or Multicrystalline

This has a slightly lower conversion efficiency compared to single crystalline but manufacturing costs are also lower. Module efficiency averages about 10% to 11%.

String Ribbon

This is a refinement of polycrystalline production, there is less work in production so costs are even lower. Module efficiency averages 7% to 8%.

Amorphous or Thin Film

Silicon material is vaporized and deposited on glass or stainless steel. The cost is lower than any other method. Module efficiency averages 5% to 7%.

Check with manufacturer for module's accurate conversion efficiency.

Photovoltaic Panels

PV panels include one or more PV modules assembled as a pre - wired, field - installable unit. The modular design of PV panels allows systems to grow as needs change. Modules of different manufacture can be intermixed without any problem, as long as all the modules have rated voltage output within 1.0 volt difference.

Photovoltaic Array

A PV Array consists of a number of individual PV modules or panels that have been wired together in a series and/or parallel to deliver the voltage and amperage a particular system requires. An array can be as small as a single pair of modules, or large enough to cover acres. 12 volt module is the industry standard for battery charging. Systems processing up to about 2000 watt - hours should be fine at 12 volts. Systems processing 2000～7000 watt - hours will function better at 24 volt. Systems running more than 7000 watt - hours should probably be running at 48 volts.

Follow the link below to see samples of complete photovoltaic - based electrical systems: Configured Solar Electric Systems.

Photovoltaic Module Performance

The performance of PV modules and arrays are generally rated according to their maximum DC power output (watts) under Standard Test Conditions (STC). Standard Test Conditions are defined by a module (cell) operating temperature of 25℃ (77 F), and incident solar irradiant level of 1000W/m^2 and under Air Mass 1.5 spectral distribution. Since these conditions are not always typical of how PV modules and arrays operate in the field, actual performance is usually 85 to 90 percent of the STC rating. Today's photovoltaic modules are extremely safe and reliable products, with minimal failure rates and projected service lifetimes of 20 to 30 years. Most major manufacturers offer warranties of twenty or more years for maintaining a high percentage of initial rated power output. When selecting PV modules, look for the product listing (UL), qualification testing and warranty information in the module manufacturer's specifications.

7.4 Reading material (Please read the article and find out the specialized vocabulary)

Photovoltaic Applications

PV has been routinely used for roadside emergency phones and many temporary construction signs, where the cost and trouble of bringing in utility power outweighs the higher initial expense of PV, and where mobile generator sets present more fueling and maintenance trouble. More than 100,000 homes in the United States, largely in rural sites, now depend on PVs as a primary power source, and this figure is growing rapidly as people begin to understand how clean and reliable this power source is, and how deeply our current energy practices are borrowing from our children. PV costs are now down to a level that makes them the clear choice not just for remote applications, but for those seeking environmentally safer solutions and independence from the ever-increasing utility power costs.

Photovoltaic Benefits

Solar power provided by photovoltaic systems lower your utility bills and insulate you from utility rate hikes and price volatility due to fluctuating energy prices.

Installing a solar system increases property value and home resale opportunities.

Purchase of a solar power system allows you to take advantage of available tax and financial incentives.

Because they don't rely on miles of exposed wires, residential PV systems are more reliable than utilities, particularly when the weather gets nasty.

PV modules have no moving parts, degrade very, very slowly, and boast a life span that isn't fully known yet, but will be measured in decades.

Solar electric systems are quiet, reliable, fossil-fuel free. Unlike mobile power generators, avoids greenhouse gas emissions.

Exercises and Discussion

1. Why do we have to encapsulate the solar cells to produce a module?
2. What is the structure of the module?
3. Please describe the encapsulation technology.

Chapter 8

PV Systems

8.1 PV system types

There are a variety of electricity generation forms for ground - based photovoltaic systems. According to the PV system whether connect to the grid or not, there are two main types which are stand alone PV system and grid connected PV system. In addition, there are also hybrid system and small - scale power system.

Specialized Vocabulary:

- PV system 光伏系统
- grid 输电网
- stand alone PV system 独立光伏发电系统
- grid connected PV system 并网光伏发电系统
- hybrid system 混合系统
- small - scale power system 小规模电源系统

8.2 Stand alone PV systems

Stand alone photovoltaic power systems are electrical power systems energized by photovoltaic panels which are independent of the utility grid. These types of systems may use solar panels only or may be used in conjunction with a diesel (or gas or petrol) generator or a wind turbine. The design of a stand alone PV - based power system is determined by the location, climate, site characteristics and equipment to be used. Figure 8 - 1 shows a schematic of a typical PV - based stand alone power system. Many countries and regions have produced standards and/or guidelines for PV systems over the last decade, and the applicable ones should always be understood and followed by designers and installers. Sometimes, compliance is a condition for subsidies or other forms of financial support.

There are two types of stand - alone PV power systems which are direct - coupled system without batteries and stand alone system with batteries (shown in Figure 8 - 2). The basic model of a direct coupled system consists of a solar panel connected directly to adirect current (DC) load. As there are no battery banks in this setup, energy is not stored and hence it is capable of powering common appliances like fans, pumps etc. only during the day. Maximum power point trackers (MPPTs) are often used to efficiently utilize the sun's energy especially for electrical loads like positive - displacement water pumps. Impedance matching is also considered as a design criterion in direct - coupled systems. In stand alone photovoltaic power systems, the electrical energy produced by the photovoltaic panels cannot always be used directly. As the demand from the load does not always equal the solar panel capacity, battery banks are generally used. The primary functions of a storage battery in a stand alone PV system are: energy storage capacity and autonomy, to store energy when there is an excess is available and to provide it when required. Voltage and current stabilization provides stable current and voltage by eradicating transients. Surge currents supply can provide surge currents to loads like motors when required.

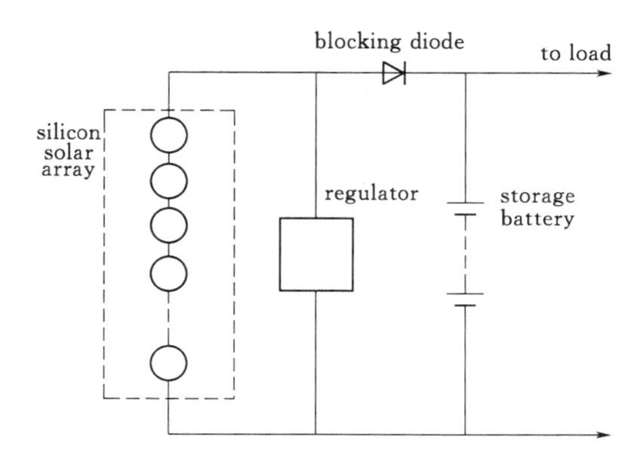

Figure 8 - 1 Simplified stand alone PV power system.
(Mack, 1979, reprinted with permission of the Telecommunication Society of Australia.)

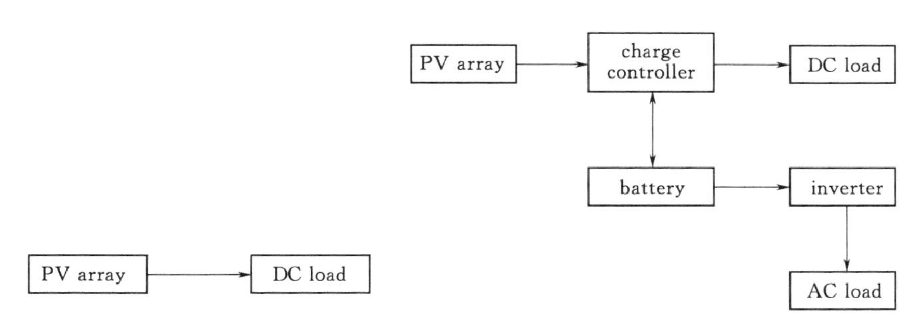

Figure 8 - 2 Direct - coupled PV system; Diagram of stand - alone PV system with battery storage powering DC and AC loads.

Specialized Vocabulary:

- direct - coupled system 直接耦合系统
- solar panel 太阳电池板
- direct current (DC) 直流
- maximum power point trackers (MPPTs) 最大功率点跟踪
- AC load 交流负载
- Inverter 逆变器

8.3 Grid connected PV systems

Grid connected photovoltaic power systems are power systems energized by photovoltaic panels which are connected to the utility grid (Figure 8 - 3) . Grid connected photovoltaic power systems consist of photovoltaic panels, MPPT, solar inverters, power conditioning units and grid connection equipment. Unlike stand alone photovoltaic power systems these systems seldom have batteries. When conditions are right, the grid connected PV system supplies the excess power, beyond consumption by the connected load, to the utility grid.

Residential grid connected photovoltaic power systems which have a relatively low capacity less than 10kW can meet the load of most consumers. They can feed excess power to the grid, which in this case acts as a battery for the system. The feedback is done through a meter to monitor power transferred. This is called net metering. Photovoltaic wattage may be less than average consumption, in which case the consumer will continue to purchase grid energy, but a lesser amount than previously. If photovoltaic wattage substantially exceeds average consumption, the energy produced by the panels will be much in excess of the demand. In this case, the excess power can yield revenue by it being sold to the grid. Depending on their agreement with their local grid energy company, the consumer only needs to pay the cost of electricity consumed less the value of electricity generated. This will be a negative number if significantly more electricity is generated than consumed. Additionally, in some cases, cash incentives are paid from the grid operator to the consumer. Connection of the photovoltaic power system can be done only through an interconnection agreement between the consumer and the utility company. The agreement details the various safety standards to be followed during the connection.

Solar energy gathered by photovoltaic solar panels, intended for delivery to a power grid, must be conditioned, or processed for use, by a grid - connected inverter. This inverter sits between the solar array and the grid, draws energy from each, and may be a large stand alone unit or may be a collection of small inverters, each physically attached to individual solar panels. The inverter must monitor grid voltage, waveform, and frequency. One reason for monitoring is if the grid is not in operation or strays too far out of its nominal specifications, the inverter must not pass any solar energy. An inverter connected to a malfunctioning power line will automatically disconnect in accordance with

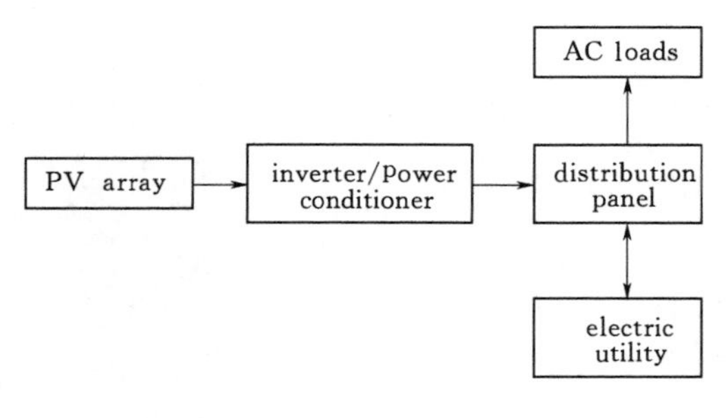

Figure 8 - 3 Grid connected PV system.

safety rules, for example UL1741[1], which vary by jurisdiction. Another reason for the inverter monitoring the grid is that for normal operation the inverter must synchronize with the grid waveform, and produce a voltage slightly higher than the grid itself, in order for energy to smoothly flow outward from the solar array.

A grid connected photovoltaic power system will reduce the power bill as it is possible to sell surplus electricity produced to the local electricity supplier. Grid connected PV systems are comparatively easier to install as they do not require a battery system. Grid interconnection of photovoltaic power generation systems has the advantage of effective utilization of generated power because there are no storage losses involved. A photovoltaic power system is carbon negative over its lifespan, as any energy produced over and above that to build the panel initially offsets the need for burning fossil fuels. Even though the sun doesn't always shine, any installation gives a reasonably predictable average reduction in carbon consumption.

Specialized Vocabulary:

- power conditioning units 功率调节单元，功率调节器
- Residential grid connected photovoltaic power systems 住宅光伏并网发电系统
- utility grid 公用电网
- fossil fuels 化石燃料
- carbon negative 碳负极
- solar array/PV array 太阳电池阵列/光伏阵列
- inverter/power conditioner 功率调节器
- distribution panel 配电盘

8.4 Hybrid systems

A hybrid system combines PV with other forms of generation, usually a diesel generator. Biogas is also used. The other form of generation may be a type able to modulate power output as a function of demand. However more than one renewable form of energy may be used e. g. wind. The photovoltaic power generation serves to reduce the consumption of non renewable fuel. Hybrid systems are the most often found on islands. Pellworm island in Germany and Kythnos island in Greece are notable examples (both are combined with wind). The Kythnos plant has reduced diesel consumption by 11.2%.

There has also been work showing that the PV penetration limit can be increased by deploying a distributed network of PV + CHP (combined heat and power) hybrid systems in the U. S. The temporal distribution of solar flux, electrical and heating require-

[1] UL1741 - *Standard for Inverters, Converters, Controllers and Interconnection System Equipment for Use with Distributed Energy Resources.*

ments for representative U. S. single family residences were analyzed and the results clearly showed that hybridizing CHP with PV can enable additional PV deployment above which is possible with a conventional centralized electric generation system. This theory was re-confirmed with numerical simulations using per second solar flux data to determine that the necessary battery backup to provide for such a hybrid system is possible with relatively small and inexpensive battery systems. In addition, large PV + CHP systems are possible for institutional buildings, which again provide backup for intermittent PV and reduce CHP runtime. Diagram of photovoltaic hybrid system is shown in Figure 8 - 4.

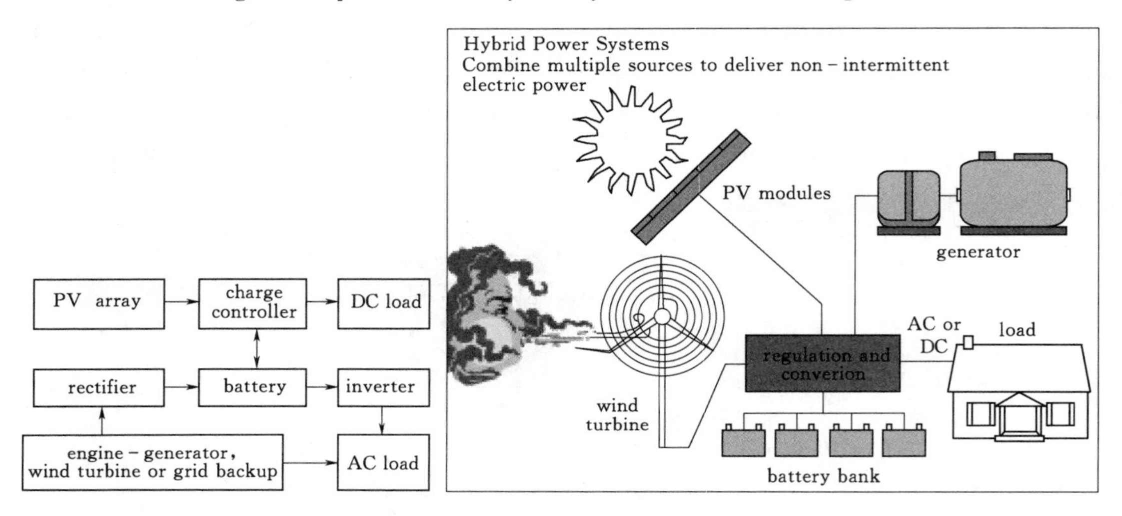

Figure 8 - 4 Diagram of photovoltaic hybrid system.

Specialized Vocabulary:

- diesel generator 柴油发电机
- biogas 生物气、沼气
- solar flux 太阳辐射通量
- centralized electric generation system 集中式发电系统
- combined heat and power 热电联供
- rectifier 整流器

8.5 Reading material (Please read the article and find out the specialized vocabulary)

Photovoltaic(PV)Systems(From Canada Mortgage and Housing Corporation)

Photovoltaic System Overview

Photovoltaic (PV) systems are used to convert sunlight into electricity. They are a safe, reliable, low-maintenance source of solar electricity that produces no on-site pollution or emissions. PV systems incur few operating costs and are easy to install on most Canadian homes. PV systems fall into two main categories—off-grid and grid-connected. The "grid" refers to the local electric utility's infrastructure that supplies electricity to homes and businesses. Off-grid systems are installed in remote locations where there is no utility grid available.

PV systems have been used effectively in Canada to provide power in remote locations for transport route signaling, navigational aids, remote homes, telecommunication, and remote sensing and monitoring. Internationally, utility grid-connected PV systems represent the majority of installations, growing at a rate of over 30% annually. In Canada, as of 2009, 90% of the capacity is in off-grid applications; however, the number of grid-connected systems continues to grow because many of the barriers to interconnection have been addressed through the adoption of harmonized standards and codes. In addition, provincial policies supporting grid interconnection of PV power have encouraged a number of building-integrated PV applications throughout Canada.

With rising electricity costs, concerns with respect to the reliability of continuous service delivery and increased environmental awareness of homeowners, the demand for residential PV systems is increasing. This *About Your House* aims to inform homeowners of what they need to consider before purchasing a system. The information presented will focus on grid-connected PV systems. To learn more about off-grid applications, consult CMHC's *Research Highlight* fact sheet Energy Use Patterns in Off-Grid Houses.

PV System Components

The most critical component of any PV system is the PV module, which is composed of a number of interconnected solar cells. PV modules are connected together into panels and arrays to meet various energy needs, as shown in Figure 8-5. The solar array is connected to an inverter that converts the Direct Current (DC) generated by the PV array into Alternating Current (AC) compatible with the electricity supplied from the grid. AC output from the inverter is connected to the home's electrical panel or utility meter, depending on the configuration. Various AC and DC disconnects are installed to ensure safety when working on the systems.

Metering

There are two different types of metering arrangements that can be used, depending on the local utility. The first is net metering, depicted in Figure 8-6. In this configuration, the utility charges you for your net consumption of electricity. When you are producing more electricity than you are consuming, your meter will essentially run backwards providing you with a credit. If you have a large system and produce a net surplus of electricity over the course of a year, utilities generally do not currently pay you for the surplus. Instead, accounts are generally reset to zero after a given period, often on a given day every year.

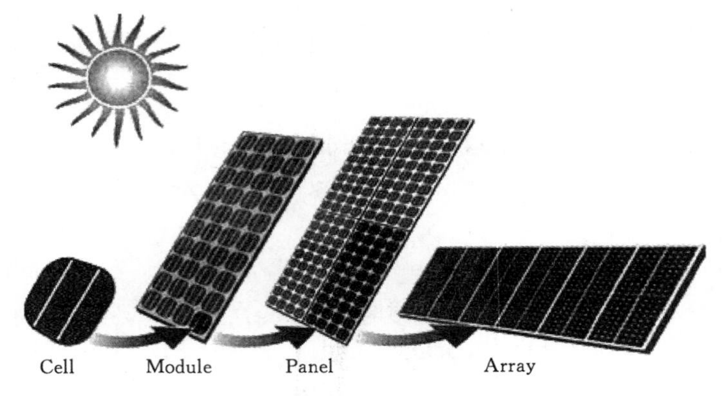

Figure 8 - 5 Components of a PV array.

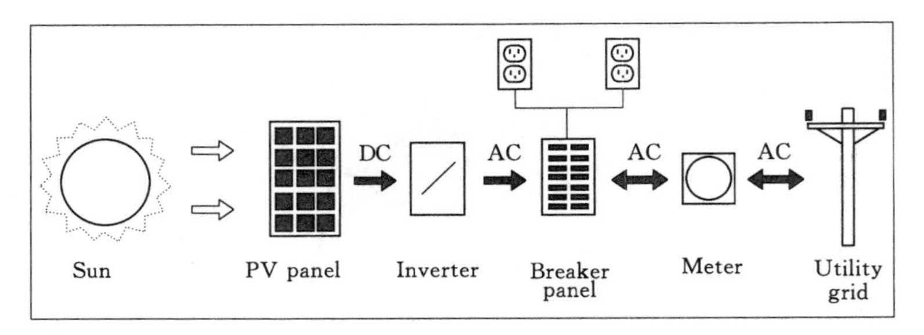

Figure 8 - 6 Net - metering PV system configuration.

The second metering arrangement is depicted in Figure 8 - 7, where the electricity generated by the PV system is measured by a separate utility meter. This metering configuration is used when the utility pays homeowners a different rate for electricity that is generated than what is taken from the grid. For example, in 2009, the Ontario provincial government started offering 20 - year fixed price contracts paying homeowners \$ 0. 802 for every kilowatt - hour produced from rooftop systems of less than 10kW[1]. These types of contracts, known as feed - in tariffs, are used to accelerate the adoption of renewable energy technologies and are discussed in more detail later.

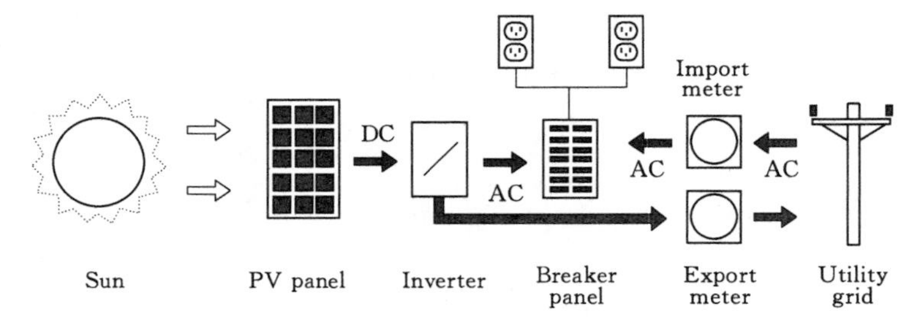

Figure 8 - 7 PV generated electricity is individually measured.

[1] Visit http://fit.powerauthority.on.ca for more information.

Backup Power

With systems configured as in Figure 8 - 6 and Figure 8 - 7, the system shuts down during power outages. In such a case, inverters are designed to sense the outage and automatically disconnect all power going to the utility meter as a safety requirement to protect utility service employees that may be working on the power lines. So even though you have a PV system, it would not be available during power outages. In order to have backup power, you need to add a battery bank. The whole domestic electrical load is too large to be entirely powered, but some inverters have the capability to continue powering an emergency sub - panel that can be used to provide power to critical loads (e. g. refrigerator, security systems, etc.) in the case of a power outage, as depicted in Figure 8 - 8. In addition to a battery bank, this configuration requires a charge controller that is able to effectively manage the batteries charging from the PV system, to ensure their optimal performance and extend their life expectancy. This system is more costly and loses some of the efficiency advantages of a battery - less system.

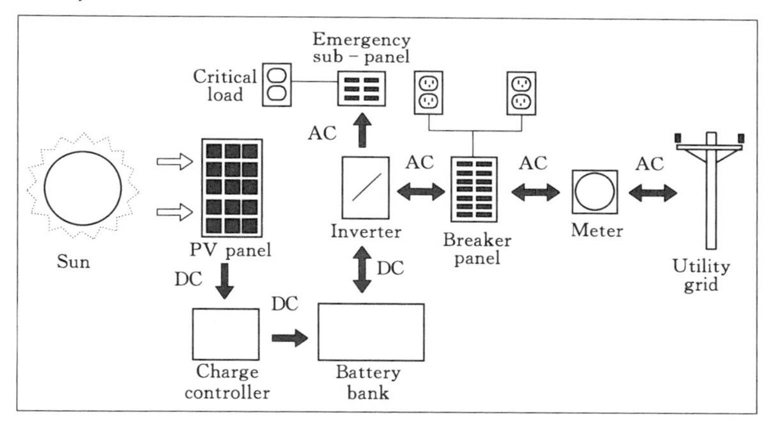

Figure 8 - 8 Net - metering PV system configuration with emergency backup.

System Design Issues

Evaluating Solar Electricity Generation Potential

It is wise to consult a PV professional at the design stage, as most dealers offer design and consultation services. Ensure that the dealer has proven experience in designing and installing the type of system you want. The Canadian Solar Industries Association (CanSIA) offers a PV Technician certificate program, and graduates have good knowledge of the design, installation and operation of home - sized PV systems. In addition, a number of community colleges across Canada have started to offer programs that cover PV system installations.

The first step in evaluating the potential of solar electricity for your home is a site assessment. PV modules are extremely sensitive to shading. Cells within a PV module and PV modules within an array are often connected in series. Think of these cells as forming a long chain, and the amount of current flowing through the chain is limited by the weakest link, i. e. the shaded cell or module. The shaded cell or module will act as a resistor. For example, if one PV module in an array of 20 modules is completely shaded, it can reduce the output power of the entire array by 100%. In addition, given that the module will be acting as a resistor stopping the current flow, it will heat up to the point where it can become damaged.

Therefore, when evaluating different locations to mount a PV array, a shading analysis needs to be performed that will identify when and where shading will occur taking into consideration that during the winter months the sun is lower in the sky and tall objects, such as trees and buildings, cast longer shadows. In most cases, the ideal location for a solar array is on the roof of the house. This alleviates most shading concerns, and its large, flat surface makes mounting relatively easy. However, chimneys and other rooftop projections need to be considered in the shading analysis. Also, the future mature height of nearby trees should be used in the evaluation instead of current tree heights.

Properly aiming modules due south with an appropriate tilt will maximize the solar energy that the PV array collects; however, small variations of up to 15° in orientation or tilt will not significantly affect performance. As a general rule, a tilt angle equal to the latitude of the site will maximize yearly performance. Reducing the tilt by 15° does not affect performance significantly (Table 8 - 1); however, a lower tilt will result in more snow accumulation in the winter. At higher angles, snow generally melts off on its own. At lower angles, snow can accumulate, reducing the power produced in the winter. However, given that most of the yearly output is produced outside winter, snow accumulation will not drastically reduce the annual performance of the system.

Table 8 - 1　　Yearly PV potential (kWh/kW) at varying tilts

All south facing	Yearly PV potential/[(kW · h) · kW^{-1}]			
	Latitude tilt - 15°	Latitude tilt	Latitude tilt +15°	Vertical, 90° tilt
Regina	1355	1361	1295	1055
Toronto	1173	1161	1095	801
Vancouver	1026	1009	939	717
St. John's	946	933	879	686

In order to assist in assessing the PV generation potential across Canada, Natural Resources Canada developed Photovoltaic potential and solar resource maps of Canada that give an estimated PV electricity production for over 3500 Canadian municipalities. The maps and tables provided present monthly and annual electricity generation per kilowatt of installed PV. As shown in Table 8 - 2, Canadian cities have a good solar potential, compared to many cities worldwide. One of our least sunny locations, St. John's, has more solar potential than cities in Germany and Japan, which are the world leading countries in solar electricity generation.

Table 8-2 Yearly PV potential of major Canadian cities and major cities worldwide

Major Canadian cities and capitals	Yearly PV potential /[(kW·h)·kW^{-1}]	Major cities worldwide	Yearly PV potential /[(kW·h)·kW^{-1}]
Regina (Saskatchewan)	1361	Cairo, Egypt	1635
Calgary (Alberta)	1292	Capetown, South Africa	1538
Winnipeg (Manitoba)	1277	New Delhi, India	1523
Edmonton (Alberta)	1245	Los Angeles, U.S.A.	1485
Ottawa (Ontario)	1198	Mexico City, Mexico	1425
Montréal (Quebec)	1185	Regina, Canada	1361
Toronto (Ontario)	1161	Sydney, Australia	1343
Fredericton (New Brunswick)	1145	Rome, Italy	1283
Québec (Quebec)	1134	Rio de Janeiro, Brazil	1253
Charlottetown (Prince Edward Island)	1095	Beijing, China	1148
Yellowknife (Northwest Territories)	1094	Washington, D.C., U.S.A.	1133
Victoria (British Columbia)	1091	Paris, France	838
Halifax (Nova Scotia)	1074	St. John's, Canada	933
Iqaluit (Nunavut)	1059	Tokyo, Japan	885
Vancouver (British Columbia)	1009	Berlin, Germany	848
Whitehorse (Yukon)	960	Moscow, Russia	803
St. John's (Newfoundland and Labrador)	933	London, England	728

Source: Natural Resources Canada. (2007). Photovoltaic potential and solar resources maps of Canada. Retrieved February 1, 2010, from https://glfc.cfsnet.nfis.org/mapserver/pv/rank.php? NEK=e.

PV System Sizing

In off-grid PV system applications, the PV array and associated battery banks must be carefully sized to be able to meet the load demands through periods with the lowest solar availability. In grid-connected applications, the presence of the grid eliminates the need to closely match the system size with the year-round electrical loads. For net-metered systems where the utility does not pay for excess electricity generation, the estimated annual solar electricity generation should be less than or equal to the annual electricity consumption as there is no financial benefit to generating more electricity than you need. For systems with a battery bank serving an emergency sub-panel, the battery bank must be sized factoring in the size of the emergency electrical loads, the PV system size, and how long emergency backup power is needed (see CMHC's About Your House: Backup Power for Your House).

Sizing of grid-connected PV systems can be approached in a number of ways depending on your objectives which could include:

- To maximize PV generation for a given budget;
- To offset your yearly purchased electricity;

- To offset a portion of your family's carbon footprint;
- To completely take advantage of available unshaded south - facing roof area;
- To reshingle a south - facing roof with PV roofing tiles;
- To improve aesthetics; and/or
- To take advantage of a government or utility incentive.

PV Panels

The three most common types of solar cells are distinguished by the type of silicon used in them: monocrystalline, polycrystalline and amorphous. Monocrystalline cells produce the most electricity per unit area and amorphous cells the least. If you want to maximize solar electricity generation for a given area, then you should select the most efficient monocrystalline PV panels you can afford. If, on the other hand, your goal is to cover a given area at the lowest cost, then you may wish to buy amorphous panels. If you are concerned with maximizing your solar electricity generation for the lowest cost, then it is best to look at the cost - effectiveness of a panel regardless of its technology by examining its cost per rated production:

$$\frac{\text{PV panel cost}}{\text{Rated PV panel output(watts)}} \Rightarrow \frac{\$}{\text{watt}}$$

For example, you want to compare the cost - effectiveness of a 160 - watt PV panel from manufacturer A selling at $ 800, to a 60 - watt PV panel from manufacturer B selling for $ 350. In this case, the more expensive panel from manufacturer A is more cost - effective at $ 5/watt compared to $ 5.83/watt for the other panel. Other factors should also be considered, such as the quality of the product. Good quality PV panels have 20 - to 25 - year warranties, have gone through testing evaluations and bear the appropriate certification labels. Also, some PV panels might be more expensive, but may also be more easily installed and thus less expensive overall. As discussed in the next section, some PV panels are designed to act as roofing tiles or shingles. Although they might be more expensive on a $ /watt basis, you also need to factor in the avoided cost of shingles or other roofing material.

Inverter Consideration

Once the PV array is sized, the size of the inverter is determined to maximize the performance of the system. If you plan to expand your PV system in the future, you may wish to oversize the inverter in order to be able to meet the additional demands of the larger system. Adequate wall space to mount the inverter and other associated components is also required in the utility room or next to your electrical panel. Small systems may only require a 0.6m × 0.9m (2ft. × 3ft.) wall area, while larger systems may require a 1.2m × 1.2m (4ft. × 4ft.) space. Some inverters are designed to withstand harsh conditions and can be mounted on an exterior wall, therefore not requiring any interior wall space. Alternatively, each PV module can be fitted with its own micro - inverter eliminating the need for one large inverter and minimizing the impacts of shading on the performance of the overall PV array.

Battery Bank

If the system has batteries, then a battery enclosure that is vented and protected against freezing will be necessary. Car batteries are not optimal for PV systems as they are designed to deliver a high current for a short period, whereas backup batteries for household applications need to deliver a relatively continuous current over extended periods. Special deep - discharge batteries are best suited. Certain types of deep - discharge batteries release small quantities of hydrogen when being charged and should be kept in a ventilated enclosure, well away from open flames or sparks. Consult your PV or battery dealer to determine the size of battery bank you need, and the installation and venting requirements for your chosen battery system.

PV System Installation

When it comes to installing PV panels on your house, there are a number of mounting options available.

Building - Integrated Products

EQuilibrium™ housing strives to achieve a balance between our housing needs and those of our natural environment. Click here to learn more about the CMHC EQuilibrium™ initiative and demonstration homes across Canada.

A number of building - integrated PV (BIPV) products are available, where the PV system essentially becomes an integral part of the building envelope. PV roofing tiles are available and were used on an EQuilibrium™ demonstration home, as shown in Figure 8 - 9.

Figure 8 - 9 Avalon Discovery 3, an EQuilibrium™ demonstration home in Red Deer, Alberta uses PV roofing tiles.

Another EQuilibrium™ home used a different option where a flexible, thin, amorphous PV panel is applied to a standing - seam metal roof, as shown in Figure 8 - 10. With that system, it is very difficult to distinguish the PV array from the metal roof.

Standard PV Panels Installed on Racking System

Standard PV panels can be mounted together on racking systems that fit on a typical roof (see Figure 8 - 11). PV systems convert 5% to 20% of the incident solar energy into electricity, a small portion is reflected, and the rest gets converted into heat. Without dissipating this heat, PV panels heat up and their efficiencies start to decrease. To address this, a

Figure 8 - 10 ÉcoTerra™, an EQuilibrium™ demonstration home in Eastman, Quebec uses amorphous PV panels stuck directly on its metal roof.

igure 8 - 11 The Now House®, an EQuilibrium™ demonstration home in Toronto, Ontario has standard PV panels mounted on its roof. Now House® is a registered trademark of the Now House Project Inc. used under license.

small air space is typically left between the PV panels and the roof to allow for air circulation to help cool the PV panels.

If you do not have sufficient roof space, a PV racking system can extend beyond your roof, like that on the EQuilibrium™ home shown in Figure 8 - 12. This configuration will experience greater wind loads, which should be considered when the system is designed.

There are a number of factors that need to be taken into consideration when designing and installing a racking system. You need to ensure that the panels are safely secured to the rack, and that the rack is safely secured

Figure 8 - 12 The Riverdale NetZero Project, an EQuilibrium™ demonstration duplex in Edmonton, Alberta has a PV racking system that extends beyond the roof.

to the roof. You may need to get your system certified by a structural engineer. Consult with your installer or your municipality to see what requirements exist in your area.

It is best to select a racking system designed for roofs and to follow the manufacturer's installation specifications. All roof penetrations for both the mounting hardware and electrical equipment need to be carefully sealed to avoid any water penetration in the future. PV systems can also be mounted vertically on a wall, but will produce less electricity, as shown in Table 8 - 1. If you do not have sufficient south - facing roof space but have a large yard, there are a number of pole - mounting options available.

If you are installing a PV system on an existing roof, you may wish to replace the existing shingles, if they have only a few years of life remaining. You do not want to have to take off the PV system shortly after its installation in order to replace the underlying roof. If you are installing a PV system on a new roof that is covered under warranty, you should ensure that adding a racking system with roof penetrations will not void your warranty. Adding a PV system on top of an existing roof can help extend its life, as the PV system will shelter the roof from the elements.

Equipment Selection

While safe installations of electrical systems are covered under the Canadian Electrical Code, the Canadian Standards Association (CSA) governs product safety. CSA has standards for all electrical components, including solar equipment and all electrical equipment must carry an approval label. Products that are purchased outside Canada may not have undergone the testing process that the same product goes through when brought in by a solar product distributor. It is possible to find good quality PV modules that meet testing standards such as IEC 61215 crystalline silicon design qualification test performance (or the IEC 61646 for thin film modules) and the IEC 61730 (or the equivalent UL 1703) safety test. In addition, inverters have to meet the CSA C22.2 standard no. 107.1 - 01 to allow their interconnection to the grid. Discuss this with your solar dealer and electrical inspector before proceeding to install these products—often a "special inspection" or extra safety measures will satisfy electrical code requirements.

It is important to remember that PV systems are modular, and can be expanded as energy needs grow or as budgets allow. It is wise to anticipate future needs by purchasing larger or oversized wires, switching gears and controls, so that these components will not have to be replaced to accommodate a larger PV system.

PV components have no moving parts—which keeps maintenance requirements to a minimum. Good quality PV modules are typically warranted for 20 to 25 years, and have life expectancies exceeding 40 years. PV panel efficiency can degrade over the years and warranties generally cover specific panel performance over the length of the warranty. The majority of the other electronic components, such as charge controls and inverters, will generally last ten or more years, if their ratings are not exceeded. Batteries typically need replacement every five to ten years.

Ask about component warranties and whether the dealer will guarantee the system. Inquire about after - sales service, including where the products need to be sent for warranty service, and who pays for shipping. An inverter that fails and needs to be shipped across the country for repairs could mean that you will be without solar power for a lengthy period—some professional dealers supply "loaners" while equipment is in for repairs.

Assuring a Safe Installation

The Canadian Electrical Code and its provincial equivalents govern installations of electrical systems. Section 50 of the Canadian Electrical Code describes the special requirements that apply to solar PV systems. In most cases, equipment must be installed by a provincially certified electrician. However, many electricians are not yet familiar with the design features of solar electrical systems and, while they will be able to install the system to meet the existing codes, they may be unable to maximize the PV system performance. In some jurisdictions, local electrical inspectors will allow non-certified electrical installers to install DC equipment, such as PV modules, charge controls and batteries. To help ensure a safe installation with optimal system performance, check whether the solar dealer has an electrician on staff or access to approved subcontractors. If contracting electricians directly, ensure that they are familiar with the design issues of PV systems by asking to see solar-related accreditation and/or a list of past PV system projects they installed, along with references.

To ensure that electrical systems, including PV systems, comply with the Electrical Code, the system may need to be inspected by the provincial Electrical Safety Authority (ESA)—your utility can supply local contact information. By having an electrical inspection done, you are ensuring the system is installed properly and is safe. Your local utility might have other requirements before allowing you to connect your PV system to the utility grid.

Integrating PV Into New House Construction

If you are in the process of designing a new house or doing major renovations, you may want to consider installing a PV system, or at least preparing your house to be "PV ready." You have an opportunity to substantially reduce costs and increase system performance. Although you may not yet be ready to invest in a PV system, the fact is that electricity prices will continue to rise while concerns about the reliability of the utility supply and the environment, combined with the decreasing cost of PV systems, will make solar electricity much more viable in Canada in the future.

While doing this preparation work, you may also wish to consider making your home "solar ready" for both PV and solar domestic hot water systems. Preparing your house to be solar ready now costs approximately $300 to $400 but can save thousands of dollars in the future. Natural Resources Canada has identified the following five basic requirements to make a home solar ready:

1. A roof location of suitable size, pitch and orientation;
2. Labelled conduits from the mechanical room to the attic area below the future PV location;
3. Extra plumbing valves and fittings on the water heater (for solar hot water systems);
4. An electrical outlet at the planned solar tank location (for solar hot water systems);

5. Construction plans that indicate the future component locations.

Orienting the house on the building lot to maximize its solar exposure and installing a roof with the correct solar pitch can maximize the performance of the PV array. Alternatively, if the lot does not permit a house to be oriented south, consider a roof shape that will have a south - facing area. Landscaping features, such as trees, should be considered when preparing the site—removing trees or moving the house site slightly can make a significant difference in available solar radiation. Remember that trees can grow a couple of feet per year and mature tree heights should be considered when determining shading potential. Although trees can have a detrimental impact on PV system performance, they can offer other benefits such as summer shading, reducing heat island effect, providing a windbreak, adding privacy, improving air quality, providing wildlife habitat that must also be considered. By carefully selecting the variety of trees and their location, you can enjoy the benefits of trees without shading your PV system.

Wires should be installed before interior walls are enclosed, as this will reduce installation time and hide unsightly conduits. Conduit runs through walls, for battery enclosure cables, battery vents, etc. , should be done at the time of construction. It is far less expensive to put conduit runs in place when installing the foundation walls than to have to drill holes later. As solar systems generate low - voltage DC power, the system wires are generally larger than normal house wiring. Minimizing the distances of wire runs is an effective method of reducing costs and increasing system efficiency.

Commissioning and Contracting with Utility

Utilities and their regulators in Canada are only beginning to address the issues of on - site generation, where individual homeowners are their own power suppliers. Discuss the status of regulations in your utility area with your local solar supplier or utility—in some cases, utilities have not yet set up a single point of contact for this new breed of customer, which can lead to delays in obtaining permission to connect to the local utility grid.

A number of currently installed standard electric meters have not been approved for net - metering applications in Canada. A more expensive electronic meter that is approved may need to be purchased. Some utilities cover the meter costs, whereas others charge the customer.

Some utilities in Canada are thinking of moving to "time - of - day" billing, which can be advantageous for homes with net - metered PV systems. This is because most solar systems generate excess electricity during peak - times—when electricity costs can be four or more times the average cost. Times when the PV system is not generating electricity and the homeowner is purchasing more electricity would typically occur more often during off - peak times, when the price of electricity is lower.

Once you obtain approval from your local utility to connect your system to the utility grid, you can turn on your system and start generating electricity. It is a good idea to compare the expected performance of the system with the actual performance, to ensure that all components are operating as expected. Keeping track of your monthly or yearly PV generation over time will help you identify problems with your system. After factoring for annual variations in solar energy, if your system is still underperforming, it may be that one of your PV panels or another component of your system is malfunctioning.

Financing and Incentive Measures

For off - grid applications, PV systems are often cost - effective as they are competing against fuel - powered generators or power line extensions, which typically cost $5000 to $10,000 per kilometre. However, for grid - connected PV systems, it is difficult to justify the installation of PV systems purely on the basis of current economics, given the current relatively low cost of grid electricity in most areas of Canada. However, some people are starting to treat PV systems like any other house upgrade. Instead of deciding whether it is cost - effective at the time of purchase, they are deciding whether they can afford it and considering their future needs along with the associated benefits of reducing one's overall environmental impact.

The current cost of PV systems ranges from $8000 to $10,000 per installed kilowatt, including all system components. In an effort to help accelerate the uptake of PV systems and drive down costs, some provinces and utilities are considering various incentive measures. As mentioned previously, one such measure is a feed - in - tariff (FIT), where a renewable energy generator is offered a premium for electricity produced, for a set term.

The most successful application of a FIT program to help accelerate the adoption of PV systems was seen in Germany, where the program helped the country become the world leader in installed PV capacity, despite its less than favourable solar resource. The first jurisdiction to offer a FIT program in North America was Ontario: it offered $0.42/(kW · h) of electricity generated from solar energy for 20 years and had 240 contracts for systems under 10kW at the beginning of 2009. Ontario recently revised its system after the passing of its Green Energy Act and has increased the rate it pays homeowners to $0.802/(kW · h) for roof - mounted systems under 10kW for at least 20 years. In 2009, this rate was higher than any other jurisdictions offering similar programs. The $0.802/(kW · h) tariff was chosen by the Ontario government based on an analysis that found that proponents could generally be expected to recover project costs and earn a reasonable rate of return at that price.

Final Thoughts on PV Systems

Energy efficiency and conservation are important measures that should be considered in conjunction with PV systems. It is far cheaper to save a kilowatt - hour than to produce one.

Even though PV systems may not be cost - effective in your area now, there is a wide variety of reasons why homeowners are considering generating some portion—if not all—of their energy requirements using PV systems. PV systems provide a buffer against rising energy prices, and the presence of an on - site battery system can supply electricity during utility power outages. Solar power can also help make a difference in the way that we address climate change and our impact on the environment.

Exercises and Discussion

1. Please describe the characteristics and differences among the stand – alone PV system, the grid – connected PV system and the hybrid system?

2. What are the applications for these three different PV systems?

References

[1] Hugo de M, Ingo H, Jose H, Philippe M, etc. R&D for PV products generating clean electricity, European roadmap for PV R&D, contact reference: ENK6 - CT2001 - 20400 (2004).

[2] Huacuz J, Urrutia M. Proceedings of the International Workshop Charge Controllers for Photovoltaic Rural Electrification Systems, Electrical Research Institute Cuernavaca, Mexico (1998).

[3] Nieuwenhout F. Monitoring and Evaluation of Solar Home Systems: Experiences with Applications of Solar PV for Households in Developing Countries, Report ECN - C - 00 - 089, Netherlands Energy Research Foundation, Patten (2000).

[4] Turton R. "Band Structure of Si: Overview", in Hull R (Ed), Properties of Crystalline Silicon, INSPEC, Stevenage, UK (1999).

[5] Green M, Keevers M. Optical properties of intrinsic silicon at 300 K. Prog Photovolt. 3, 189 - 192 (1995).

[6] Kolodinski S, Werner J, Wittchen T, Queisser H. Quantum efficiencies exceeding unity due to impact ionization in silicon solar cells. Appl. Phys. Lett. 63, 2405 - 2407 (1993).

[7] Clugston D, Basore P. Modelling Free - Carrier Absorption in Solar Cells. Prog. Photovolt. 5, 229 - 236 (1997).

[8] Sproul A, Green M. Improved value for the silicon intrinsic carrier concentration from 275 to 375 K. J. Appl. Phys. 70, 846 - 854 (1991).

[9] Wholesale Solar - Three Photovoltaic Technologies: Polycrystalline and thin film. http://www.wholesalesolar.com/Information - SolarFolder/celltypes.html.

[10] The Solar Plan - Converting Photons to Electrons. http://www.thesolarplan.com/articles/how - do - solar - panels - work.html.

[11] Phillips J, Birkmire R, McCandless B, Meyers P, Shafarman W. Polycrystalline heterojunction solar cells: A device perspective. Phys. Stat. Sol. (b) 194, 31 - 39 (1996).

[12] Fan J, Palm B. Optimal design of amorphous/crystalline tandem cells. Sol. Cells. 11, 247 - 261 (1984).

[13] Nell M, Barnett A. The spectral p - n junction model for tandem solar - cell design. IEEE Trans. Elec. Dev. ED - 34, 257 - 265 (1987).

[14] Contreras M, Egaas B, Ramanathan K, Hiltner J, Swartzlander A, Hasoon F, Noufi R. Progress toward 20% efficiency in Cu (In, Ga) Se_2 polycrystalline thin - film solar cells. Prog. Photovolt. 7, 311 - 316 (1999).

[15] Tanaka Y, Akema N, Morishita T, Kushiya K. Proc. 17th Euro. Conf. Photovoltaic Solar Energy Conversion, 989 - 994 (2001).

[16] Wieting R. CIS product introduction: Progress and challenges. AIP Conf. Proc. 462, 3 - 8 (1999).

[17] Burgess R, Chen W, Devaney W, Doyle D, Kim N, Stanbery B. Electron and proton radiation effects on GaAs and CuInSe/sub 2/ thin film solar cells. Proc. 20th IEEE Photovoltaic Specialist Conf., 909 - 912 (1988).

[18] Jasenek A, Rau U, Weinert K, Kotschau I, Werner J. Radiation resistance of Cu (In, Ga) Se_2 solar cells under 1 - MeV electron irradiation. Thin Solid Films, 387, 228 - 230 (2001).

[19] Rocheleau R, Meakin J, Birkmire R, Proc. 19th IEEE Photovoltaic Specialist Conf. , 972 - 976 (1987).

[20] Mitchell K, Eberspacher C, Ermer J, Pauls K, Pier D. CuInSe/sub2/cells and modules. IEEE Trans. Electron. Devices 37, 410 - 417 (1990).

[21] Kazmerski L, Wagner S, "Cu - Ternary Chalcopyrite Solar Cells", in Coutts T, Meakin J, Eds, Current Topics in Photovoltaics, 41 - 109, Academic Press, London (1985).

[22] Haneman D. Properties and applications of copper indium diselenide. Crit. Rev. Solid State Mater. Sci. 14, 377 - 413 (1988).

[23] Rockett A, Birkmire R. $CuInSe_2$ for photovoltaic applications. J. Appl. Phys. 70, R81 - R97 (1991).

[24] Rockett A, Bodegard M, Granath K, Stolt L. Na incorporation and diffusion in CuIn/sub 1 - x/Ga/sub x/Se/sub 2/. Proc. 25th IEEE Photovoltaic Specialist Conf. , 985 - 987 (1996).

[25] Rau U, Schock H. Electronic properties of Cu (In, Ga) Se_2 heterojunction solar cells - recent achievements, current understanding, and future challenges. Appl. Phys. A69, 131 - 147 (1999).

[26] Alferov Zh I, Andreev V M, Kagan M B, Protasov Ⅱ I and Trofim V G. "Solar - energy converters based on p - n $Al_xGa_{1-x}As$ - GaAs heterojunctions," Fiz. Tekh. Poluprovodn. 4, 2378 (1970) [Sov. Phys. Semicond. 4, 2047 (1971)].

[27] Yablonovitch E, Miller O D, Kurtz S R. "The opto - electronic physics that broke the efficiency limit in solar cells" . 38th IEEE Photovoltaic Specialists Conference, p. 001556, doi: 10. 1109/PVSC. 2012. 6317891, ISBN 978 - 1 - 4673 - 0066 - 7, (2012).

[28] Schnitzer I, et al. "Ultrahigh spontaneous emission quantum efficiency, 99. 7% internally and 72% externally, from AlGaAs/GaAs/AlGaAs double heterostructures" . Applied Physics Letters 62 (2): 131 (1993).

[29] Wang X, et al. "Design of GaAs Solar Cells Operating Close to the Shockley - Queisser Limit". IEEE Journal of Photovoltaics 3 (2): 737 (2013).

[30] Pulfrey L D. Photovoltaic Power Generation. New York: Van Nostrand Reinhold Co. ISBN 9780442266400, (1978).

[31] Rivers P N. Leading edge research in solar energy. Nova Science Publishers, ISBN 1600213367, (2007).

[32] Hoppe H and Sariciftci N S. "Organic solar cells: An overview" . J. Mater. Res. 19 (7): 1924 - 1945 (2004).

[33] Halls J J M, Friend R H, Archer M D, Hill R D. Clean electricity from photovoltaics. London: Imperial College Press, pp. 377 - 445, ISBN 1860941613, (2001).

[34] Chiu S W, Lin LY, Lin H W, Chen Y H, Huang Z Y, Lin Y T, Lin F, Liu Y H, Wong K T. "A donor - acceptor - acceptor molecule for vacuum - processed organic solar cells with a power conversion efficiency of 6. 4%" . Chemical Communications 48. doi: 10. 1039/C2CC16390J. Retrieved 16 January (2014).

[35] Li B, et al. "Review of recent progress in solid - state dye - sensitized solar cells" . Solar Energy Materials and Solar Cells 90 (5): 549 - 573 (2006).

[36] Yella A, Lee H W, Tsao H N, et al. Porphyrin - sensitized solar cells with cobalt (Ⅱ/Ⅲ) - based redox electrolyte exceed 12 percent efficiency [J] . Science, 334 (6056): 629 - 634 (2011).

[37] Freitag M, Teuscher J, Saygili Y, Zhang X, etc. Dye - sensitized solar cells for efficient power generation under ambient lighting. Nat. Photon. doi: 10. 1038/nphoton. 2017. 60 (2017).

[38] Lee M M, Teuscher J, Miyasaka T, Murakami T N, Snaith H J. Efficient hybrid solar cells based

on meso - superstructured organometal halide perovskites. Science 338, 643 - 647 (2012).

[39] Kojima A, Teshima K, Shirai Y, Miyasaka T. Organometal halide perovskites as visible - light sensitizers for photovoltaic cells. J. Am. Chem. Soc. 131, 6050 - 6051 (2009).

[40] Noh J H, Im S H, Heo J H, Mandal T N, Seok S. Chemical management for colorful, efficient, and stable inorganic - organic hybrid nanostructured solar cells. Nano Lett. 13, 1764 - 1769 (2013).

[41] Sheng R, Ho - Baillie A W Y, Huang S J, Cheng Y B, Green M A. Four - terminal tandem solar cells using $CH_3NH_3PbBr_3$ by spectrum splitting. JPCL 6, 3931 - 3934 (2015).

[42] Heo J H, et al. Efficient inorganic - organic hybrid heterojunction solar cells containing perovskite compound and polymeric hole conductors. Nature Photon. 7, 486 - 491 (2013).

[43] Burschka J, et al. Sequential deposition as a route to high - performance perovskite - sensitized solar cells. Nature 499, 316 - 319 (2013).

[44] Liu M, Johnston M B, Snaith H J. Efficient planar heterojunction perovskite solar cells by vapor deposition. Nature 501, 395 - 398 (2013).

[45] Malinkiewicz O, et al. Perovskite solar cells employing organic charge - transport layers. Nature Photon. 8, 128 - 132 (2014).

[46] Ball J M, Lee M M, Hey A, Snaith H J. Low - temperature processed meso - superstructured to thin - film solar cells. Energy Environ. Sci. 6, 1739 - 1743 (2013).

[47] "Stand -Alone Photovoltaic Systems" . http: //renewableenergy. com/. Retrieved 2011 - 07 - 21.

[48] Abujasser A. "A STAND - ALONE PHOTOVOLTAIC SYSTEM, CASE STUDY: A RESIDENCE IN GAZA" . trisanita. org. Retrieved 2011 - 07 - 21.

[49] "SBatteries and Charge Control in Stand - Alone Photovoltaic Systems - Fundamentals and Application" . http: //localenergy. org. Retrieved 2011 - 07 - 21.

[50] Elhodeiby A S, Metwally H M B, Farahat M A, "Performance anyalysis of 3. 6kW rooftop grid connected photovoltaic system in EGYP" . International Conference on Energy Systems and Technologies (ICEST 2011): 151 - 157, Retrieved 2011 - 07 - 21.

[51] "Grid Connected Solar Electric - Photovoltaic (PV) Systems " . http: //powernaturally. org. Retrieved 2011 - 07 - 21.

[52] "Grid -connected photovoltaic system" . http: //soe - townsville. org. Retrieved 2011 - 07 - 21.

[53] "International Guideline For The Certification Of Photovoltaic System Components and Grid - Connected Systems" . http: //iea - pvps. org. Retrieved 2011 - 07 - 21.

[54] PV resources website, Hybrid power station accessed 10 Feb 2008.

[55] Daten und Fakten at the Wayback Machine, Pellworm island website, in German. (archived July 19, 2011).

[56] Darul'a I, Stefan M. "Large scale integration of renewable electricity production into the grids". Journal of Electrical Engineering 58 (1): 58 - 60 (2007).

[57] Pearce J M. "Expanding Photovoltaic Penetration with Residential Distributed Generation from Hybrid Solar Photovoltaic + Combined Heat and Power Systems" . Energy 34: 1947 - 1954 (2009).

[58] Derewonko P, and Pearce J M. "Optimizing Design of Household Scale Hybrid Solar Photovoltaic+Combined Heat and Power Systems for Ontario", Photovoltaic Specialists Conference (PVSC), 34th IEEE, pp. 1274 - 1279 (2009).

[59] Mostofi M, Nosrat A H, and Pearce J M. "Institutional - Scale Operational Symbiosis of Photovoltaic and Cogeneration Energy Systems" International Journal of Environmental Science and Technology 8 (1), pp. 31 - 44 (2011).